The Acheulian Site of Gesher Benot Ya'aqov

Volume II

Vertebrate Paleobiology and Paleoanthropology Series

Edited by

Eric Delson
Vertebrate Paleontology, American Museum of Natural History,
New York, NY 10024, USA
delson@amnh.org

Eric J. Sargis
Anthropology, Yale University
New Haven, CT 06520, USA
eric.sargis@yale.edu

Focal topics for volumes in the series will include systematic paleontology of all vertebrates (from agnathans to humans), phylogeny reconstruction, functional morphology, Paleolithic archaeology, taphonomy, geochronology, historical biogeography, and biostratigraphy. Other fields (e.g., paleoclimatology, paleoecology, ancient DNA, total organismal community structure) may be considered if the volume theme emphasizes paleobiology (or archaeology). Fields such as modeling of physical processes, genetic methodology, nonvertebrates or neontology are out of our scope.

Volumes in the series may either be monographic treatments (including unpublished but fully revised dissertations) or edited collections, especially those focusing on problem-oriented issues, with multidisciplinary coverage where possible.

For other titles published in this series, go to
www.springer.com/series/6978

The Acheulian Site of Gesher Benot Ya'aqov

Volume II

Ancient Flames and Controlled Use of Fire

Nira Alperson-Afil

Institute of Archaeology, The Hebrew University of Jerusalem, Jerusalem, Israel

Naama Goren-Inbar

Institute of Archaeology, The Hebrew University of Jerusalem, Jerusalem, Israel

Springer

Nira Alperson-Afil
Institute of Archaeology
The Hebrew University of Jerusalem
Jerusalem
Israel
alperson@mscc.huji.ac.il

Naama Goren-Inbar
Institute of Archaeology
The Hebrew University of Jerusalem
Jerusalem
Israel
goren@cc.huji.ac.il

ISBN 978-90-481-3764-0 e-ISBN 978-90-481-3765-7
DOI 10.1007/978-90-481-3765-7
Springer Dordrecht Heidelberg London New York

Library of Congress Control Number: 2010921299

Cover illustration: An artist's view of fireplaces on the margins of the paleo-Lake Hula. Illustration by Amir Balaban

Printed on acid-free paper

Springer is part of Springer Science+Business Media (www.springer.com)

A Volume in the Gesher Benot Ya‘aqov Subseries

Coordinated by

Naama Goren-Inbar

Institute of Archaeology, The Hebrew University of Jerusalem

For us nothing is more common than fire; but man could have wandered in the desert for millions of years without once having seen fire on earthly soil. Let us grant him an erupting volcano, a forest set on fire by lightning; hardened in his nakedness against the rigors of the seasons, would he have run forward at once to warm himself? Would he not rather have taken flight?

(Bachelard 1938:23)

Foreword

A View from Western Europe

Most archaeologists would agree that the emergence of stone tool manufacture and the management of fire are the two most significant events in the cultural evolution of early humans. The oldest known stone artifacts are securely dated to 2.6–2.5 Ma at several localities in Ethiopia; their association with ungulate remains and observations of cut marks prove that one of their main functions was for butchery (Domínguez-Rodrigo et al. 2005). The record of early stone tools from a number of sites in the time span 2.5–2.0 Ma is unequivocal; tool use and manufacture were a regular activity with evidence of planning, foresight and considerable technical skills (Delagnes and Roche 2005). In contrast, the timing of the human control of fire is not fully resolved and the antiquity of its habitual use has been debated until now.

This book provides very strong evidence of the habitual use of fire by early humans at the Acheulian site of Gesher Benot Ya'aqov (Israel). The sedimentary sequence at the site is 34 m thick, and it represents different depositional environments, mainly beaches along the margins of a paleo-lake. The Matuyama-Brunhes chron boundary, dated to 0.78 Ma, occurs in the lower part of the sequence. The 15 archaeological levels discussed in the book occur above this boundary and clearly indicate repeated occupations over a long span of time; the lowermost occupation level occurs 4 m above the Matuyama-Brunhes boundary, and the highest is 13 m above the boundary. The duration of the entire depositional sequence at GBY is estimated as ca. 100 kyr.

The evidence is strong because it is based on different kinds of data, in particular the fact that burned microartifacts (≤2 cm) occur in localized concentrations in many superimposed archaeological levels. Similar approaches to the spatial distribution of lithic artifacts and burned ecofacts as a way to locate "invisible" hearths have been used by archaeologists working on Mesolithic open-air sites in NW Europe (Sergant et al. 2006). At many of these open-air sites, hearths are "invisible" because they are not stone-built, charcoal and ash remains have disappeared as a result of postdepositional processes, such as wind and rain, and reddening of the soil did not occur due to a lack or low amount of iron in the sandy soils. The presence of hearths is revealed by the spatial clustering of burned items, that is lithics and burned hazelnut shells. When research on the use of fire and "phantom hearths" started at the GBY site, those approaches were unknown to the Israeli scholars; the convergence on the use of spatial analyses to locate invisible structures at sites of very different ages is a comment on the adequacy of the methodology used in this book.

More importantly, of the 15 assemblages analyzed in this book, five contain clusters with frequencies of burned microartifacts in the order of 3.7–5.8%. As noted by Alperson-Afil and Goren-Inbar in Chapter 4, these values are comparable to frequencies of burned microartifacts 1–3 cm in size at Magdalenian sites in Western Europe, in particular at two sites, Hauterives-Champréveyres and Monruz, both located on the shores of the Neuchâtel Lake in Switzerland and dated at around 13,000 BP. At these sites, the hearths (40 at Monruz, 11 at Hauterives-Champréveyres) are exceptionally well preserved; they are very visible structures with heated slabs, charcoal and burned bones (Leesch 1997; Bullinger et al. 2006). The spatial analysis of microartifacts at these two sites was done explicitly with the purpose of demonstrating the significance of burned microartifacts for indicating the actual location of an "invisible" hearth, better than the distribution of macroartifacts, which are often rejected away from the combustion area.

The review of the earliest sites with putative evidence of fire (Chapter 1) shows that the evidence (charcoal, heat-altered sediments, burned stones, burned bones and ash) is often fragmentary and judged insufficient. The strongest claims are from Member 3 of Swartkrans (dated to 1.9–1.65 Ma) and Koobi Fora (site FxJj 20, dated to 1.5 Ma). At those sites the heating of bones was supported by ESR analyses of bone and TL analyses of reddened sediments, respectively. Since the lithic assemblages of GBY demonstrate the introduction of African biface-making techniques into Eurasia, the authors argue that fire-making too may reflect an African tradition and a wave of human migration out of Africa (Section 4.5).

Did control of fire play a role in the colonization of Europe?

The colonization of Europe, especially of the regions where temperatures at times dropped below the freezing point, is generally tied to the use of fire. Yet evidence for fire in the Early and early Middle Pleistocene is extremely weak or more exactly negative until about 400 ka. The review of early European sites provided in Chapter 1 shows that good evidence of fire (burned artifacts dated by the TL method, and burned bones and patches of reddened earth, interpreted as remnants of fireplaces) comes only from two sites dated to MIS 11, i.e. about 400 ka, Beeches Pit in England and Schöningen in Germany. At Terra Amata (France), in addition to artifacts dated by TL, there was one clear charcoal concentration; the charcoal was identified as *Pinus sylvestris*. The age estimates of the site vary between 230 ± 40 ka and 380 ± 80 ka (Villa 1983; Falguères et al. 1988). By MIS 7 and 6, several other sites provide evidence of the use of fire, although visible fireplaces were often not preserved: e.g. Vaufrey, La Cotte de St. Brelade, and Orgnac. At some sites overlapping palimpsests of fireplaces formed large combustion areas (Bau de l'Aubesier, Grotte XVI).

Of direct relevance to European prehistory are sites that do *not* have traces of fire, yet might be expected to have such evidence. In Table i I present the current state of our knowledge in Western Europe and include early sites that do not have evidence of fire. In addition, I include evidence of fire originating from sites younger than those discussed in Chapter 1, thus complementing the review of early European sites.

I have excluded many open-air occurrences in fluviatile or clearly disturbed contexts and some sites with ambiguous or underreported evidence. Burned flint artifacts have provided TL dates at two sites: Biache St. Vaast (175 ± 13 ka, mean age of layer IIA; Tuffreau and Sommé 1988) and Maastricht-Belvedere (ca. 250 ka for Unit IV C; Roebroeks 1988) but I have no further information. Two Spanish sites that seem to be of Late Matuyama age, Cueva Negra (a rock shelter in the Estrecho del Rio Quípar, Murcia province; www.um.es/antropofisica/english/cuevanegra.html) and La Boella near Tarragona (open-air site; www.diaridetarragona.com/) are not included because current investigations are too preliminary. By MIS 4 and 3, Mousterian sites that have evidence of fire (concentrations of charcoal, burned bones, stone-lining) are numerous (I count at least 20 in France), so they are not listed in the table except for cases of stone-lined fireplaces, which are uncommon occurrences prior to the Upper Paleolithic.

Table i shows that evidence for the use of fire in the earliest European record, prior to 400 ka, is lacking. It can be argued that sites such as those in the Orce region, Isernia and Venosa Notarchirico have been affected by water transport and that Boxgrove may represent brief occupations and butchery episodes. Eight charcoal particles have been found at Boxgrove, and one charcoal fragment was found in a layer above the main occupation level. Clearly we

Table i Fire and fireplaces in Pleistocene Western Europe. Abbreviations. – = unreported; ? = unlikely or unclear or insufficiently described; yes = reported in some detail, sometime confirmed by TL dates on burned flint; no = reported as absent; A = charcoal analyzed by a botanist. Site type: O, open air; C, cave; E, enclosed (doline, rock shelter). Hearths = structured, spatially defined features presenting a concentration of burned items (ash, charcoal, heated slabs, reddened sediment, burned bones, burned artifacts); one attribute alone may be insufficient, it is the combination or the spatial concentration of at least some attributes that is significant

Site	Site type and reference no.	Age	Charcoal	Burned stones or artifacts	Burned bones	Burned sediments	Hearths	Evidence of fire
Fuentenueva 3, Barranco Léon (Orce)	O (1)	Late Matuyama, probably 1.3/1.2 Ma	–	–	–	–	–	–
Sima del Elefante (Atapuerca)	C (1)	> 800 ka?	–	–	–	–	–	–
Gran Dolina, TD 6 (Atapuerca) (24)	C (1–3)	800 ka	–	–	No	–	No	No
Isernia (central Italy)	O (4)	606 ± 2 ka $^{40}Ar/^{39}Ar$	–	–	??	??	No	No
Venosa Notarchirico (southern Italy)	O (4)	640 ± 70 ka on tephra by TL on quartz grains	–	–	No	–	No	No
Boxgrove (England) (21)	O (5)	MIS 13 Ca. 500 ka	A Dispersed	–	No	–	No	No
Arago; Layers D-Q (southern France)	C (6, 7)	> 350 ka MIS 12–14	No	No	No	–	No	No
Schöningen 13 I (Germany)	O cf. Chapter 1 and (8)	MIS 11 Ca. 400 ka	–	Artifacts (TL dates)	–	Yes	No	Yes
Beeches Pit (England)	O cf. Chapter 1	MIS 11	–	Artifacts (TL dates)	Yes	Yes	No	Yes
Vértesszőllős (Hungary)	O cf. Chapter 1	MIS 9/11	–	–	In concentrations	Patches	No	To be verified
Bilzingsleben (Germany)	O cf. Chapter 1	MIS 9/11	Yes	Yes	Yes	–	–	Likely
Barnham (England)	O (9)	MIS 11	A Dispersed	5 natural pcs (TL)	–	–	No	Not in situ
Terra Amata (SE France)	O cf. Chapter 1 (10, 11)	380 ± 80 ka (ESR) 230 ± 40 ka (TL)	A Localized	Artifacts (TL dates)	Yes	?	Flat lens	Yes
Orgnac 3; Layers 2, 6 (southern France)	E (12,13)	MIS 9–8	Ashes observed during excavation	–	In concentrations	–	?	Yes
Lunel Viel (southern France)	C (14)	MIS 9?	Dispersed	Burned stones	No	?	?	Not in situ
Cagny l'Epinette (Somme, northern France)	O (12)	MIS 9	–	–	Some	–	No	?
Vaufrey Layer VIII (Dordogne, France)	C (15)	MIS 7	Yes Localized	No	Some	No	No	Yes
La Cotte de St. Brelade (Jersey Island)	E cf. Chapter 1	MIS 7 and 6	Yes	Yes	Yes	Yes	Not preserved	Yes

Table i (continued)

Site	Site type and reference no.	Age	Charcoal	Burned stones or artifacts	Burned bones	Burned sediments	Hearths	Evidence of fire
Menez Dregan (Brittany), layer 5	C (16)	MIS 7, ca. 200 ka, TL on quartz grains and artifacts	A	Yes	?	Yes	?	Yes
Bau de l'Aubesier (SE France) Layer H-1	E (17)	191 ± 15; 169 ± 17 ka TL on artifacts	Charcoal and ashes	Yes	Yes	Yes	5 m^2 combustion area, 40 cm thick	Yes
Bolomor, Layer XIII and younger Layers XI, IV, II (Spain)	C (18,19,20)	MIS 7, 6 and 5e, various TL dates	Ashes in layer II	Artifacts	Yes, in layer IV	Yes	3 in layer IV, 2 with stones in layer XIII	Yes
Lazaret Cave (SE France)	C (7)	MIS 6	A localized	–	Yes	–	Flat conc.	Yes
Bau de l'Aubesier Layer IV (SE France)	E (17)	MIS 5	Yes	Yes	Yes	Yes	55 m^2 combustion area, 20 cm thick	Yes
Vaufrey Layer IV (Dordogne, France)	C (21)	MIS 5; 120 ± 10 ka, TL on artifacts	A. Burned plant material	Artifacts	Yes	Yes	Not preserved	Yes
Les Canalettes (southern France)	E (22)	Ca 70 ka	A Charcoal and coal	Stones, TL on flint artifacts	?	Yes	One stone-lined	Yes
Grotte XVI, layer C (Dordogne, France)	C (23,24)	ca 60 ka, TL dates on flint	Altered ash derived from wood and grass	Yes	Yes	Yes	Overlapping palimpsests of hearths forming a combustion zone >12 m^2	Yes
Vilas Ruivas (Portugal)	O (25)	50–60 ka	–	Burned stones	–	–	Two, stone-lined	Yes
La Combette, layer D (SE France)	E (26)	MIS 3	Charcoal	Burned stones	–	Yes	Four, flat and stone-lined	Yes
Abric Romaní (Spain)	E (27,28)	MIS 3	Yes	Yes, TL	Yes	?	Many, some stone-lined	Yes

(1) Santonja and Villa 2006; (2) Díez et al. 1999; (3) Fernández-Jalvo et al. 1999; (4) Villa and Lenoir in press; (5) Roberts and Parfitt 1999; (6) Falguères et al. 2004; (7) de Lumley 2006; (8) Thieme 2000; (9) Ashton et al. 1998; (10) Falguères et al. 1988; (11) Villa 1991; (12) Moigne and Barsky 1999; (13) Moncel et al. 2005; (14) Le Grand 1994; (15) Rigaud and Geneste 1988; (16) Mercier et al. 2004; (17) Lebel and Trinkaus 2002; (18) Blasco López 2006; (19) Fernández-Peris 2007; (20) Sanchis Serra and Fernández-Peris 2008; (21) Courty 1988; (22) Meignen 1993; (23) Karkanas et al. 2002; (24) Rigaud et al. 1995; (25) Vega Toscano et al. 1999; (26) Texier et al. 1998; (27) Pastó et al. 2000; (28) Vaquero and Pastó 2001

cannot exclude natural fires. At High Lodge, also dated to MIS 13 like Boxgrove (Ashton et al. 1992), five charcoal particles were found but they were dispersed in the deposits. No burned bones and no burned artifacts have been reported from either Boxgrove or High Lodge. Flecks of charcoal were also found at Swanscombe and Hoxne, dated to MIS 11 (Wymer 1999), but again they were dispersed in the sediments and could have been the result of natural fires; there is no evidence of burned artifacts either.

But what about occupation sites in caves which in later times have often provided striking evidence of fire, such as Bau de l'Aubesier, Grotte XVI, Lazaret and Middle Paleolithic/Middle Stone age caves in Israel and in South Africa?

Traces of fire have been found in the upper part of the sequence at Arago, in layers younger than 350 ka, but no charcoal, no burned bones nor any other evidence of fire have been reported from the lower levels of Arago (dated to MIS 12–14). This is surprising because taphonomic analyses have been carried out (e.g., Moigne and Barsky 1999), and there are paleontological papers and doctoral theses on specific taxa (e.g., Monchot 1996); faunal and lithic remains are very abundant.

No burned bones or burned artifacts have been reported from Gran Dolina, layer TD6. Rare charcoal particles have been found in micromorphological slides, but the origin of the sediments is from the exterior of the cave, and there is evidence of low energy transport (Valleverdú et al. 2001); thus the charcoal may not be in situ. However, the high density of human, faunal and lithic remains, and their state of preservation and refitting (Díez et al. 1999; Fernández-Jalvo et al. 1999) clearly indicate an occupation in situ with little postdepositional disturbance. In sum, both at Gran Dolina TD6 and at Arago, layers D to Q, this absence of evidence of fire is in need of an explanation.

I have suggested in the past (Villa and Bon 2002) that absence or non-systematic use of fire may be one of the reasons why the settlement of Europe took a rather long time. Prior to 400,000 years ago the total number of sites is quite small, and this suggests rather sporadic and discontinuous settlement patterns. Only from MIS 11 onward does the utilization of fire become a significant feature of the record.

I think now that the evidence from GBY should encourage European archaeologists to take a closer look at their data, in particular microartifacts, and to investigate taphonomic and diagenetic processes that may explain the disappearance of fire traces. In the absence of such detailed studies, explanations for the absence of fire at the Early and early Middle Pleistocene European sites would be flawed and may be short-lived.

Paola Villa
University of Colorado Museum
UCB 265, Bruce Curtis Building
Boulder, Colorado, 80309-0265, USA
and
UMR 5199-PACEA
Institut de Préhistoire et Géologie du Quaternaire
Université Bordeaux 1
Avenue des Facultés
33405 Talence, France

Preface

The discovery of evidence for fire at Gesher Benot Ya'aqov was not part of our expectations from the outset; nor was the study of fire, its control and its cultural implications initially among the many and diverse goals of the project. The discovery illustrates the fascination and unpredictability of the archaeological discipline. The presence of fire at the site, and its occurrence in all of the prehistoric occupations revealed during seven field seasons, turned it into a major research objective. The results of this research are presented in this volume.

The origin of the research lies in the proximity of the Gesher Benot Ya'aqov Acheulian site to the Jordan River. This resource was exploited for the wet-sieving of all the sediments removed during excavation. The apparatus was constructed in such a way that the excavators could sit on small stools in the river and operate hanging sieves of 2 mm mesh, which were submerged in water. All sieved material larger than 2 mm was washed, dried, and later sorted. It was during this sorting process that the flint microartifacts that form the bulk of the database of this study were collected and later analyzed.

The small lithic component could not be identified during excavation, due to the waterlogged nature of the sediments, the dark color of the deposit, and the necessity to shade the excavated surface from the sun and moisten it continuously to preserve the organic materials (wood, bark, fruits and seeds) embedded in it. Thus, the recovery of burned flint microartifacts during sieving in the field was accidental, and was later verified in the lab at Kibbutz Gadot, where the expedition was lodged throughout the field seasons. This fortuitous discovery led to a prolonged study of the evidence for fire in all of the archaeological horizons of the Gesher Benot Ya'aqov excavations.

The identification and sorting of microartifacts of all raw materials was carried out at the Institute of Archaeology of the Hebrew University of Jerusalem, a procedure that necessitated the involvement of many individuals. The sorting, which lasted from 1989 to 2007, was carried out by students; most of them had no previous experience in archaeology and came from different departments of the Faculty of Humanities, School of Law, School of Education and Faculty of Social Sciences.

The broken hearts and many other non-archaeological issues that were discussed while tweezers and brushes were operated could have been the subject of an extensive sociological study in themselves. We achieved the sorting of over half a million microartifacts, and the children of the first sorters will probably appear as students of the Hebrew University very shortly.

The order and magnitude of the task we planned made some of the funding agencies very skeptical about the feasibility of the proposed research. One perceived disadvantage was the lack of similar attempts, though they are widespread nowadays. Clearly, the task of sorting needed perseverance more than anything else. Important changes took place throughout the years of sorting and analysis. For example, the GIS and other program packages developed tremendously. The first attempts to explore the applicability of GIS programs to the distribution of microartifacts were rejected by experts, due to lack of experience in intra-site projects and the overwhelming size of the database.

We have carried out this task with a deep sense of duty and with constant curiosity and anticipation. Indeed, every archaeological excavation brings with it the obligations of recovering, recording and preserving, which are all components of the attempt to reconstruct ancient cultures and past ways of life. At prehistoric archaeological sites, where we rarely encounter constructed features (not to mention monumental structures or historical records), we must endeavor to make the most of the data retrieved. Throughout the course of this study we were guided by the concept of *structures latentes*, first established by Leroi-Gourhan. This concept recognizes the fact that the archaeological record conceals information that is not visible at first sight, since it does not exhibit directly observable features. Accordingly, ancient fireplaces were embedded within the archaeological levels at GBY, though they lacked apparent color, constructed contour or clear accumulations of ashes and burned material. Their presence could be discerned only through careful examination of spatial patterns, particularly those of the small lithic items.

The use of the *structures latentes* concept at GBY enabled the remarkable discovery of Acheulian hearths. Moreover, the fact that such hearths are recorded throughout the long archaeological sequence suggests that fire was not only used but *controlled* by the Acheulian hominins of GBY as early as 0.79 million years ago. Conclusions like these, and their implications for the archaeological, anthropological and evolutionary sciences, illustrate the great potential of such studies. For us, despite the immense amount of time and resources required to accomplish the task, this long journey was truly worthwhile, as it enabled us to recognize an exceptionally significant aspect of the lives and behavior of the GBY hominins.

Jerusalem, February 2009

Nira Alperson-Afil
Naama Goren-Inbar

Acknowledgments

The realization of this volume, an almost impossible task that could not have been accomplished by a single person, is the result of the work of many individuals and an enormous number of working hours. Despite the attractions of the research subject, the amount of work necessary to accomplish the study put off more than a student or two. While the task initially seemed easy due to the small number of burned flint items, with progress it became evident that each of the archaeological horizons encompassed these artifacts and that the job waiting to be done was enormous.

The fieldwork at Gesher Benot Ya'aqov was supported over the years by the Leakey Foundation, the National Geographic and the Hebrew University of Jerusalem. The laboratory work was supported over the years by the Leakey Foundation, the Irene Levi-Sala CARE Archaeological Foundation, the Israel Science Foundation (several grants) and the Hebrew University of Jerusalem. Part of the work was carried out by the Center of Excellence, supported by the Israel Science Foundation (grant No. 300/06). The study of burned flint microartifacts was supported by grants from the German–Israeli Foundation (GIF I-896–208.4/2005) and the Israel Science Foundation (grant No. 886/02). We also wish to thank the Faculty of Humanities, Hebrew University of Jerusalem, and the Ruth Amiran Fund of the Institute of Archaeology, Hebrew University of Jerusalem, for their support of the publication of this volume.

The authors wish to thank the Gesher Benot Ya'aqov staff and students at the Institute of Archaeology, the Hebrew University of Jerusalem, for sorting of sediments and microartifacts throughout the years (1989–2007): Tchiya Alon, Doron Ben-Ami, Rachel Berman, Orly Boyum, Keren Braverman, Anat Cohen, Taufik Daa'dalla, Doron Dag, Hila Debono, Olga Dobovski, Michal Dovrat, Tomer Dror, Irit Eckshtein, Caroline Felus, Keren Finkelshtein, Talia Goldman, Emma Goldshmit, Natalya Gubenko, Debby Hochman, Ronit Israeli, Sharon Kaplan, Keren Katzir, Karni Klein, Christopher Konlan, Irena Laschiver, Liora Levy, Ravit Lin, Amit Marari, Maya Margalit, Efrat Mashiach, Noa Mashiach, Zinovi Matzkevitch, Sigalit Mazugi, Michal Meir, Uri Milshtein, Hadas Motro, Tali Nevo, Boaz Ofek, Dan Oriyan, Oded Ramati, Dan Reshef, Menachem Rogel, Keren Rosenberg, Karmit Rubin, Eli Shabo, Reut Shalib, Vivian Shalom, Ruthy Shimron, Iris Shmila, Esther Sivan, Guy Stiebel, Hagar Taub, Lilli-Shantal Tepber, Sigalit Tzaidi, Smadar Tzaidi, Eyal Vadai and Yulya Volodosky.

We would like to express our deep gratitude to two individuals who consistently follow and support our work and assist in bringing it to publication in a highly professional manner. Noah Lichtinger contributed her expertise in graphics and was involved in each of the many minute details and decisions of color and layout. Sue Gorodetsky edited the text in her usual meticulous and efficient way.

We wish to express our thanks to Yuval Goren, treasurer of the Israeli Prehistoric Society, and to the administrative staff of the Institute of Archaeology of the Hebrew University, Frida Lederman and Benny Sekay, who, usually unseen and working behind the scenes, greatly assisted us in completing the present study.

Many thanks are due to Eric Delson, editor of the *Vertebrate Paleobiology and Paleoanthropology* book series, who generously accepted our study for publication and improved it with expertise and dedication. We thank the editorial staff at Springer, particularly Tamara Welschot and Judith Terpos, for their constant interest and efficiency. Eric Delson, Craig Feibel, Paola Villa, Richard Wrangham, and an anonymous reviewer read and commented on earlier versions of this manuscript. We are very thankful for, and most appreciative of, their scholarly remarks and comments. Paola Villa has also agreed to make a contribution from her vast knowledge and take upon herself the task of writing the foreword. We are most thankful for her input.

Nira Alperson-Afil would like to express her gratitude to the following. I thank the Hebrew University of Jerusalem for scholarships granted during work on the Ph.D. thesis on which this manuscript is based, and the supportive members of the Ph.D. committee: Prof. Anna Belfer-Cohen, Prof. Isaac Gilead and Prof. Erella Hovers. I thank and cherish Prof. Naama Goren-Inbar for her enthusiasm, encouragement and endless support, which brought this study to its completion. I could not hope for a better supervisor and teacher. Special thanks are due to Prof. Ilan Sharon, Adi Ben-Nun and Prof. Isaac Gilead, who greatly assisted in constructing the methodology of this study. The stuff and students of the prehistory department at the Institute of Archaeology, particularly Prof. Anna Belfer-Cohen, Prof. Nigel Goring-Morris, Prof. Erella Hovers, Dr. Leore Grosmann, Dr. Gonen Sharon, Irena Laschiver, Hila Ashkenazi, Michal Birkenfeld, Doron Dag and Angela Davidzon, are thanked for their constant interest and encouragement. Thanks to Noah Lichtinger, Ruhama Bonfil-Piperno, Sue Gorodetsky, Rivka Rabinovitch and Hanna Fluck for their assistance and support. And finally, my love and gratitude to Hagar, Tamara, Yonatan and Evyatar.

Naama Goren-Inbar thanks Nira Alperson-Afil for her ability to integrate fragmented ideas, free associations, a general research orientation, (very) long planning and formulating it all into a coherent excellent study. She successfully carried it out with deep understanding of the anthropological background coupled with modern usage of spatial and statistical applications. Without her talent, intelligence and perseverance, this study could not have been published.

Contents

1 Introduction 1

1.1 The Early Evidence 1
1.1.1 Africa 2
1.1.2 The Levant 4
1.1.3 Asia 5
1.1.4 Europe 6
1.1.5 Summary 8
1.2 Identifying Fire: Methods and Techniques 8
1.2.1 Burned Sediments 9
1.2.2 Burned Bones 10
1.2.3 Burned Stones 10
1.3 The Acheulian Site of Gesher Benot Ya'aqov 11
1.3.1 The Depositional Sequence 11
1.3.2 Chronology 13
1.3.3 Cultural and Behavioral Issues 14

2 Framework of Research 19

2.1 Research Objectives 19
2.2 Hearth-Related Spatial Patterning 19
2.2.1 Small In Situ Refuse 20
2.2.2 Phantom Hearths 21
2.3 Research Hypotheses 22
2.3.1 Taphonomic Considerations 22
2.3.2 The Effects of Fire on Flint 23
2.3.3 Patterns of Anthropogenic Fire 24
2.4 Methodology 24
2.4.1 Excavation Methods and Provenance Recording 24
2.4.2 Analyzed Samples 25
2.4.3 Lithic Analyses 26
2.4.4 Database Construction 26
2.4.5 Generating Distribution Maps and Density Maps 28
2.4.6 Analysis of Spatial Patterns 30

3 Results 33

3.1 Layer V-5 33
3.2 Layer V-6 36
3.3 Layer I-4 39
3.4 Layer I-5 40
3.5 Layer II-2/3 41
3.6 Layer II-5 41

3.7 Layer II-5/6 45
3.8 Layer II-6 L-1 47
3.9 Layer II-6 L-2 50
3.10 Layer II-6 L-3 52
3.11 Layer II-6 L-4 55
3.12 Layer II-6 L-4b 58
3.13 Layer II-6 L-5 61
3.14 Layer II-6 L-6 64
3.15 Layer II-6 L-7 66
3.15.1 Layer II-6 L-7: Upper Occupational Floor 66
3.15.2 Layer II-6 L-7: Test Pits 69

4 Discussion and Conclusions 73

4.1 Fire Ecology 75
4.1.1 Ignition and Combustion 75
4.1.2 Heat Transfer 76
4.1.3 Fire Behavior 77
4.1.4 Probability of a Natural Fire at GBY 77
4.2 Use of Fire at GBY 79
4.2.1 Differentiation in the Distribution of Burned and Unburned Flint 79
4.3 Patterns of Human Activities and Fire at GBY 83
4.3.1 The Carcass-Processing Site of Layer V-6 83
4.3.2 The Elephant-Butchering Site of Layer II-6 L-1 87
4.3.3 The Basalt Biface Workshop Site of Layer II-6 L-4 88
4.3.4 Humans and Fire at GBY: Summary 90
4.4 Paleolithic Fire in Perspective 92
4.5 Conclusions 96

Appendices 99

Appendix 1: Burned and Unburned Flint Microartifacts 99
Appendix 2: Working Methodology for ArcMap GIS Files 100
Appendix 3A: The use of different search radii and cell sizes in kernel density maps. 104
Appendix 3B: Density maps based on data interpolation of sub-square precision 105
Appendix 4: Probability of a Natural Wildfire at GBY 106

References 107

Subject Index 115

Index of Sites 119

List of Tables

Table 2.1 The archaeological layers included in the study 25

Table 3.1 Lithic assemblage of Layer V-5 34
Table 3.2 Lithic assemblage of Layer V-6 36
Table 3.3 Lithic assemblage of Layer I-4, total assemblage (South and North areas) 39
Table 3.4 Lithic assemblage of Layer I-4 (North), spatially presented 39
Table 3.5 Lithic assemblage of Layer I-5, total assemblage (South and North areas) 40
Table 3.6 Lithic assemblage of Layer I-5 (North), spatially presented 40
Table 3.7 Lithic assemblage of Layer II-2/3 41
Table 3.8 Lithic assemblage of Layer II-5 42
Table 3.9 Lithic assemblage of Layer II-5/6 45
Table 3.10 Lithic assemblage of Layer II-6 L-1 48
Table 3.11 Lithic assemblage of Layer II-6 L-2 50
Table 3.12 Lithic assemblage of Layer II-6 L-3 53
Table 3.13 Lithic assemblage of Layer II-6 L-4 55
Table 3.14 Lithic assemblage of Layer II-6 L-4b 58
Table 3.15 Lithic assemblage of Layer II-6 L-5 61
Table 3.16 Lithic assemblage of Layer II-6 L-6 64
Table 3.17 Lithic assemblage of Layer II-6 L-7 (upper occupational floor) 67
Table 3.18 Lithic assemblage of Layer II-6 L-7 (Northern Pit) 70
Table 3.19 Lithic assemblage of Layer II-6 L-7 (Southern Pit) 70

Table 4.1 Summary table – characteristics of the high-density kernels of burned clusters 74
Table 4.2 Counts of large burned flint items according to their proximity to high-density concentrations of burned flint microartifacts in different layers of GBY 74

Table 4.3 Attributes of the GBY layers relative to the degree of spatial differentiation between burned and unburned flint microartifacts.............. 80

Table 4.4 Layer V-6: characteristics of the general bone assemblage and of the bones in proximity to the concentrations of burned flint microartifacts 86

List of Figures

Fig. 1.1 Chronological and geographical distribution of the major occurrences of early fire mentioned in the text 9

Fig. 1.2 Location map of the upper Jordan Valley (the Hula Valley and vicinity); the location of the GBY archaeological site is marked on the map 12

Fig. 1.3 Location map of archaeological excavation areas and geological trenches of GBY 13

Fig. 1.4 The northern face of Trench II; the major stratigraphic units of Area B are marked on the section 14

Fig. 1.5 General view of the excavations at Area B (1996 season); view to the south from Trench II 15

Fig. 1.6 Composite section of the GBY sequence with correlation to major stratigraphical units and cyclicity of the sequence 15

Fig. 2.1 The stages of building distribution and density maps, demonstrated on the assemblage of burned flint microartifacts from Layer II-6 L-1 (N = 754) 29

Fig. 3.1 Percentages of flint microartifacts per excavated unit in Layer V-5. (**a**) Burned flint microartifacts (N = 560) and (**b**) unburned flint microartifacts (N = 30,315) 34

Fig. 3.2 Kernel density maps of flint microartifacts in Layer V-5. (**a**) Burned flint microartifacts (N = 560) and (**b**) unburned flint microartifacts (N = 30,315) 35

Fig. 3.3 Kernel density maps of three randomly selected data sets (N = 560) for Layer V-5 35

Fig. 3.4 The kernel density map of burned flint microartifacts of Layer V-5 and excavated units in which the observed percentage of burning exceeds the expected percentage; significant SR values are marked on the map 36

Fig. 3.5 The kernel density map of burned flint microartifacts of Layer V-5 and the distribution of large burned flint items 36

Fig. 3.6 Percentages of flint microartifacts per excavated unit in Layer V-6. (**a**) Burned flint microartifacts (N = 83) and (**b**) unburned flint microartifacts (N = 4,424) 37

Fig. 3.7 Kernel density maps of flint microartifacts in Layer V-6. (**a**) Burned flint microartifacts (N = 83) and (**b**) unburned flint microartifacts (N = 4,424) 38

Fig. 3.8 Kernel density maps of three randomly selected data sets (N = 83) for Layer V-6 38

Fig. 3.9 The kernel density map of burned flint microartifacts of Layer V-6 and excavated units in which the observed percentage of burning exceeds the expected percentage; significant SR value is marked on the map 39

Fig. 3.10 The kernel density map of burned flint microartifacts of Layer V-6 and the distribution of large burned flint items 39

Fig. 3.11 Percentages of flint microartifacts per excavated unit in Layer I-4 (North). (**a**) Burned flint microartifacts (N = 40) and (**b**) unburned flint microartifacts (N = 6,457) 40

Fig. 3.12 Percentages of flint microartifacts per excavated unit in Layer I-5 (North). (**a**) Burned flint microartifacts (N = 121) and (**b**) unburned flint microartifacts (N = 14,894) 41

Fig. 3.13 Percentages of flint microartifacts per excavated unit in Layer II-2/3. (**a**) Burned flint microartifacts (N = 51); location of burned flint macroartifact is marked on the map and (**b**) unburned flint microartifacts (N = 7,323) 42

Fig. 3.14 Percentages of flint microartifacts per excavated unit in Layer II-5. (**a**) Burned flint microartifacts (N = 162) and (**b**) unburned flint microartifacts (N = 3,453) 43

Fig. 3.15 Kernel density maps of flint microartifacts in Layer II-5. (**a**) Burned flint microartifacts (N = 162) and (**b**) unburned flint microartifacts (N = 3,453) 43

Fig. 3.16 Kernel density maps of three randomly selected data sets (N = 162) for Layer II-5 44

Fig. 3.17 The kernel density map of burned flint microartifacts of Layer II-5 and excavated units in which the observed percentage of burning exceeds the expected percentage 44

Fig. 3.18 The kernel density map of burned flint microartifacts of Layer II-5 and the distribution of large burned flint items 44

Fig. 3.19 Percentages of flint microartifacts per excavated unit in Layer II-5/6. (**a**) Burned flint microartifacts (N = 400) and (**b**) unburned flint microartifacts (N = 8,487) 45

Fig. 3.20 Kernel density maps of flint microartifacts in Layer II-5/6. (**a**) Burned flint microartifacts (N = 400) and (**b**) unburned flint microartifacts (N = 8,487) 46

Fig. 3.21 Kernel density maps of three randomly selected data sets (N = 400) for Layer II-5/6 46

Fig. 3.22 The kernel density map of burned flint microartifacts of Layer II-5/6 and excavated units in which the observed percentage of burning exceeds the expected percentage; significant SR values are marked on the map 47

Fig. 3.23 The kernel density map of burned flint microartifacts of Layer II-5/6 and the distribution of large burned flint items 47

Fig. 3.24 Percentages of flint microartifacts per excavated unit in Layer II-6 L-1. (**a**) Burned flint microartifacts (N = 754) and (**b**) unburned flint microartifacts (N = 53,081) 48

Fig. 3.25 Kernel density maps of flint microartifacts in Layer II-6 L-1. (**a**) Burned flint microartifacts (N = 754) and (**b**) unburned flint microartifacts (N = 53,081) 49

Fig. 3.26 Kernel density maps of three randomly selected data sets (N = 754) for Layer II-6 L-1 49

Fig. 3.27 The kernel density map of burned flint microartifacts of Layer II-6 L-1 and excavated units in which the observed percentage of burning exceeds the expected percentage; significant SR values are marked on the map 50

Fig. 3.28 The kernel density map of burned flint microartifacts of Layer II-6 L-1 and the distribution of large burned flint items 50

Fig. 3.29 Percentages of flint microartifacts per excavated unit in Layer II-6 L-2. (**a**) Burned flint microartifacts (N = 563) and (**b**) unburned flint microartifacts (N = 73,064) 51

Fig. 3.30 Kernel density maps of flint microartifacts in Layer II-6 L-2. (**a**) Burned flint microartifacts (N = 563) and (**b**) unburned flint microartifacts (N = 73,064) 51

Fig. 3.31 Kernel density maps of three randomly selected data sets (N = 563) for Layer II-6 L-2 52

Fig. 3.32 The kernel density map of burned flint microartifacts of Layer II-6 L-2 and excavated units in which the observed percentage of burning exceeds the expected percentage; significant SR value is marked on the map 52

Fig. 3.33 The kernel density map of burned flint microartifacts of Layer II-6 L-2 and the distribution of large burned flint items 52

Fig. 3.34 Percentages of flint microartifacts per excavated unit in Layer II-6 L-3. (**a**) Burned flint microartifacts (N = 877) and (**b**) unburned flint microartifacts (N = 91,050) 53

Fig. 3.35 Kernel density maps of flint microartifacts in Layer II-6 L-3. (**a**) Burned flint microartifacts (N = 877) and (**b**) unburned flint microartifacts (N = 91,050) 54

Fig. 3.36 Kernel density maps of three randomly selected data sets (N = 877) for Layer II-6 L-3 54

Fig. 3.37 The kernel density map of burned flint microartifacts of Layer II-6 L-3 and excavated units in which the observed percentage of burning exceeds the expected percentage; significant SR values are marked on the map 55

Fig. 3.38 The kernel density map of burned flint microartifacts of Layer II-6 L-3 and the distribution of large burned flint items 55

Fig. 3.39 Percentages of flint microartifacts per excavated unit in Layer II-6 L-4. (**a**) Burned flint microartifacts (N = 1,406) and (**b**) unburned flint microartifacts (N = 105,022) 56

Fig. 3.40 Kernel density maps of flint microartifacts in Layer II-6 L-4. (**a**) Burned flint microartifacts (N = 1,406) and (**b**) unburned flint microartifacts (N = 105,022) 56

Fig. 3.41 Kernel density maps of three randomly selected data sets (N = 1,406) for Layer II-6 L-4 57

Fig. 3.42 The kernel density map of burned flint microartifacts of Layer II-6 L-4 and excavated units in which the observed percentage of burning exceeds the expected percentage; significant SR values are marked on the map 58

Fig. 3.43 The kernel density map of burned flint microartifacts of Layer II-6 L-4 and the distribution of large burned flint items 58

Fig. 3.44 Percentages of flint microartifacts per excavated unit in Layer II-6 L-4b. (**a**) Burned flint microartifacts (N = 443) and (**b**) unburned flint microartifacts (N = 7,182) 59

Fig. 3.45 Kernel density maps of flint microartifacts in Layer II-6 L-4b. (**a**) Burned flint microartifacts (N = 443) and (**b**) unburned flint microartifacts (N = 7,182) 59

Fig. 3.46 Kernel density maps of three randomly selected data sets (N = 443) for Layer II-6 L-4b 60

Fig. 3.47 The kernel density map of burned flint microartifacts of Layer II-6 L-4b and excavated units in which the observed percentage of burning exceeds the expected percentage; significant SR values are marked on the map 60

Fig. 3.48 The kernel density map of burned flint microartifacts of Layer II-6 L-4b and the distribution of large burned flint items 61

Fig. 3.49 Percentages of flint microartifacts per excavated unit in Layer II-6 L-5. (**a**) Burned flint microartifacts (N = 1,211) and (**b**) unburned flint microartifacts (N = 31,862) 62

Fig. 3.50 Kernel density maps of flint microartifacts in Layer II-6 L-5. (**a**) Burned flint microartifacts (N = 1,211) and (**b**) unburned flint microartifacts (N = 31,862) 62

Fig. 3.51 Kernel density maps of three randomly selected data sets (N = 1,211) for Layer II-6 L-5 63

Fig. 3.52 The kernel density map of burned flint microartifacts of Layer II-6 L-5 and excavated units in which the observed percentage of burning exceeds the expected percentage; significant SR values are marked on the map 63

Fig. 3.53 The kernel density map of burned flint microartifacts of Layer II-6 L-5 and the distribution of large burned flint items 63

Fig. 3.54 Percentages of flint microartifacts per excavated unit in Layer II-6 L-6. (**a**) Burned flint microartifacts (N = 251) and (**b**) unburned flint microartifacts (N = 12,282) 64

Fig. 3.55 Kernel density maps of flint microartifacts in Layer II-6 L-6. (**a**) Burned flint microartifacts (N = 251) and (**b**) unburned flint microartifacts (N = 12,282) 65

Fig. 3.56 Kernel density maps of three randomly selected data sets (N = 251) for Layer II-6 L-6 65

Fig. 3.57 The kernel density map of burned flint microartifacts of Layer II-6 L-6 and excavated units in which the observed percentage of burning exceeds the expected percentage; SR values are marked on the map 66

Fig. 3.58 The kernel density map of burned flint microartifacts of Layer II-6 L-6 and the distribution of large burned flint items 66

Fig. 3.59 Percentages of flint microartifacts per excavated unit in Layer II-6 L-7 upper occupational floor. (**a**) Burned flint microartifacts (N = 677) and (**b**) unburned flint microartifacts (N = 23,455) 67

Fig. 3.60 Kernel density maps of flint microartifacts in Layer II-6 L-7 upper occupational floor. (**a**) Burned flint microartifacts (N = 677) and (**b**) unburned flint microartifacts (N = 23,455) 68

Fig. 3.61 Kernel density maps of three randomly selected data sets (N = 677) for Layer II-6 L-7 upper occupational floor 68

Fig. 3.62 The kernel density map of burned flint microartifacts of Layer II-6 L-7 upper occupational floor and excavated units in which the observed percentage of burning exceeds the expected percentage; SR values are marked on the map 69

Fig. 3.63 The kernel density map of burned flint microartifacts of Layer II-6 L-7 upper occupational floor and the distribution of large burned flint items 69

Fig. 3.64 Percentages of flint microartifacts per excavated unit in Layer II-6 L-7 Northern Pit. (**a**) Burned flint microartifacts (N = 295); distribution of burned flint macroartifacts is marked on the map and (**b**) unburned flint microartifacts (N = 10,940) 70

Fig. 3.65 Percentages of flint microartifacts per excavated unit in Layer II-6 L-7 Southern Pit. (**a**) Burned flint microartifacts (N = 241), location of burned flint macroartifact is marked on the map and (**b**) unburned flint microartifacts (N = 6,228) 71

Fig. 4.1 General view of the excavations of Layer V-6 84

Fig. 4.2 Layer V-6. (**a**) Kernel density map of basalt microartifacts and the distribution of basalt macroartifacts and (**b**) kernel density map of unburned flint microartifacts and the distribution of flint macroartifacts 85

Fig. 4.3 Layer V-6. The field map superimposed on the kernel density map of burned flint microartifacts 86

Fig. 4.4 Layer II-6 L-1. (**a**) The exposed living floor during field work and (**b**) the elephant skull after retrieval from the field 87

Fig. 4.5 Layer II-6 L-1. The field map superimposed on the kernel density map of burned flint microartifacts 88

Fig. 4.6 Layer II-6 L-4. View of the northern part of the excavated living floor during field work 89

Fig. 4.7 Layer II-6 L-4. The field map superimposed on the kernel density map of burned flint microartifacts 90

Fig. 4.8 Layer II-6 L-2. (**a**) The complete field map superimposed on the kernel density map of burned flint microartifacts and (**b**) selected items of the field map (largest basalt slab, bones, wood, handaxes, and cleavers) and the pitted anvils superimposed on the kernel density map of burned flint microartifacts 91

Fig. 4.9 Layer II-6 L-6. (**a**) The complete field map superimposed on the kernel density map of burned flint microartifacts and (**b**) selected items of the field map (largest basalt cores and bifacial tools) and the pitted anvils superimposed on the kernel density map of burned flint microartifacts 91

Fig. 4.10 The sequential occurrence of phantom hearths in Layer II-6 by level, from topmost level (II-6 L-1) to the lowermost level (II-6 L-7 upper occupational floor) 97

Chapter 1
Introduction

The manipulation of fire was clearly a turning point for our ancient ancestors. Their ability to "domesticate" this powerful tool, lacking in any other creature, has provided us with the valuable gift of fire, whose possession has remained exclusively human ever since. As fire conferred varied advantages for early man, providing warmth and light, protection from predators, and the ability to exploit a new range of foods, the issue of human mastery of the use of fire has occupied numerous archaeological and anthropological studies (e.g., Harrison 1954; Oakley 1956; Stewart 1956; Perlès 1977; Clark and Harris 1985; Goudsblom 1986; James 1989; Olive and Taborin 1989; Wrangham et al. 1999; Villa 2001), while the question of *when* humans obtained and controlled fire has remained obscure.

Evidence from the Early and early Middle Pleistocene site of Gesher Benot Ya'aqov provides a unique opportunity to examine this controversial issue. The site, located on the shores of the paleo-Lake Hula in the Levantine Corridor, displays a variety of evidence suggesting that hominins repeatedly occupied the site for some 100,000 years (Goren-Inbar et al. 2000; Feibel 2001). Several multidisciplinary studies indicate that the hominins of Gesher Benot Ya'aqov skillfully modified stone tools, systematically butchered and consumed animal carcasses, and collected a vast range of plant foods, identified due to unique conditions of preservation in the waterlogged environment of the site (e.g., Goren-Inbar et al. 1994; Goren-Inbar and Saragusti 1996; Goren-Inbar et al. 2002a, b; Ashkenazi et al. 2005; Goren-Inbar and Sharon 2006; Rabinovich et al. 2008).

During the early excavations of the site, evidence of burning was observed by Stekelis, who reported the discovery of a burned tibia and burned bone splinters within the Acheulian deposits of Gesher Benot Ya'aqov (Stekelis 1960). The renewed excavations at the site (1989–1997) have unearthed burned flint, wood, fruits, and grains as well as charcoal fragments, which were collected from various archaeological horizons throughout the stratigraphic sequence.

This study focuses on *flint* microartifacts (≤2 cm) and macroartifacts (>2 cm). Since the exposure of flint to high temperatures (i.e., 350–500°C) results in distinctive thermal macrofractures, burned flint items (items including micro- and macroartifacts as well as natural pebbles) can easily be identified by simple visual observation. Analysis of the presence of burned flint items, their spatial setting, and the circumstances in which they were deposited in the archaeological horizons at Gesher Benot Ya'aqov constitute the fundamental objectives of this research.

The basic assumption advocated in this study is that the presence and the spatial clustering of burned flint items provide direct evidence for the use of fire. This assumption draws on a variety of ethnographic, archaeological, and ethnoarchaeological studies, which generally suggest that small burned items and their spatial arrangement are significant spatial indicators of the locations of ancient hearths. Using various procedures of spatial display and analysis, the distribution patterns of the burned and unburned flint items are examined to detect possible clusters of burned material. These clusters are interpreted in this study as remnants of hearths, indicative of the use of fire.

The burned flint items from Gesher Benot Ya'aqov may establish evidence for the use of fire as early as 790,000 years ago. This evidence contributes to the growing available data on early indications for the use of fire by early humans, thoroughly discussed in this chapter.

1.1 The Early Evidence

The fact that early hominins, unlike other animals that fear the sight of fire, were able to "domesticate" fire suggests that this stage in human evolution may symbolize more than anything else the birth of our humanity. Nonetheless, the question of when humans came to obtain and control fire remains controversial.

Attempts to evaluate this point in time are fundamentally an archaeological challenge. However, review of the archaeological data demonstrates that to a large extent the early evidence is fragmentary and inconclusive.

Commenting on James's (1989) review of the early indications of fire use, both Dennell (1989) and McGrew (1989) call attention to the fact that the evidence for fire is both direct

N. Alperson-Afil and N. Goren-Inbar, *The Acheulian Site of Gesher Benot Ya'aqov Volume II: Ancient Flames and Controlled Use of Fire*, Vertebrate Paleobiology and Paleoanthropology, DOI 10.1007/978-90-481-3765-7_1, © Springer Science+Business Media B.V. 2010

and indirect, and we should hence be cautious when using criteria of unequal value. However, it is our view that, given the antiquity of some of the evidence and its diverse nature, the use of a large range of criteria is essential when attempting to identify early use of fire and to construct a comprehensive description of the antiquity of control of fire.

Excavations of archaeological sites embedded in Pleistocene sediments have yielded various indications of fire. While a great deal has been written on the significance and implications of the acquisition of fire, the evidence can be compiled from the reports of various excavated sites. In some cases researchers were aware of the importance of their findings and conducted further analyses on the "fire-altered" features and items. In other cases, the reports merely mention the discovery of a burned item or feature without further analysis or interpretation, thus overlooking its major implications. Thus, it is important to note that, due to the large number of researchers and the diversity of methods involved in the analyses of burned artifacts and features, the discussions of early fire are extremely varied and uneven, ranging from purely descriptive to highly interpretive.

The following review attempts to integrate the evidence for the use of fire prior to the Middle Paleolithic, assembling the available evidence from the published data in as detailed a form as possible. The review follows the evidence geographically and chronologically, with the aim of tracing the time span in which this valuable technological invention emerged and spread. Following this review of the early evidence, a thorough discussion of the methodologies applied to the various types of evidence is presented.

1.1.1 Africa

The anthropological evidence from which our early evolutionary history is reconstructed originates from Africa, and it is in this continent that the earliest known evidence for the use of fire is recorded. During the 1920s, a discovery in the Makapansgat valley, Central Transvaal, suggested that fire was used as early as the times of *Australopithecus*. In 1925, pieces of bone breccia with fragments of charred bones were collected during quarrying for lime. These were given to Prof. Raymond Dart, who suspected that the breccia was a cave deposit containing hearths. Analyses of the bone fragments have suggested that these were burned (as quoted in Oakley 1956). Some 20 years later, newly discovered fossil bones of *Australopithecus* appeared to be of the same breccia as that of Makapansgat. Dart then named the "fire making" hominin *Australopithecus prometheus*.[1] However, analysis of these breccias failed to show the presence of burned particles (Oakley 1956). *Australopithecus prometheus* is thus not the earliest hominin for which the use of fire is recorded; rather, such evidence is principally recorded in Early Pleistocene sites attributed to *Homo erectus*.[2]

Early Pleistocene deposits were studied in a large-scale excavation at *Koobi Fora* in northern Kenya. The silty flood plain sediments along the western face of the Karari Escarpment were dated to ca. 1.5 Ma (Isaac and Harris 1978; Clark and Harris 1985) and yielded several archaeological sites. Of these, in the site of FxJj 20 two archaeological localities (FxJj 20 East and FxJj 20 Main) exhibiting possible evidence for fire were unearthed. Four patches of presumably burned sediments were observed near the base of the archaeological horizon at FxJj 20 East. These measured ca. 30–40 cm in diameter and were 10–15 cm thick (Clark and Harris 1985). Three patches consisted of a blocky consolidated mass of sandy silts, with a slight reddish/orange color and some flecks of stronger red. The fourth patch consisted of sandy silt sediments with a blackened zone showing an intense gray/black color, associated in part with calcification.

Samples of these patches were analyzed by thermal demagnetization, which concluded that some of the patches show strong indications of having been heated by fire to temperatures of 200–400°C (Clark and Harris 1985). In addition, thermoluminescence (TL) analysis of the samples demonstrated that the reddish patches were indeed heated more recently than the surrounding tuffs: "If there had been a general grass or forest fire, the geological TL response of the tuff would have been reduced to the same level as that of the reddish spots" (Rowlett 2000:200).

Further analysis attempted to distinguish whether these ancient fires were the result of fireplaces or burned trees. For this purpose, a study of the phytoliths originating from the reddish patches was carried out. In the case of a burned tree, a homogenous composition of phytoliths is expected. However, the samples displayed a heterogeneity of phytoliths that extends the observed heterogeneity of experimental fireplaces with deliberately mixed fuels (Rowlett et al. 1999; Rowlett 2000). The phytholith study also enabled the identification of palm wood as one of the fuels used in construction of the fireplace (Rowlett et al. 1999; Rowlett 2000).

At FxJj 20 East, additional evidence of fire is found in some "...thermally altered stone artifacts" (Clark and Harris 1985:12; unspecified raw material). These are black or reddish-orange artifacts, interpreted as discolored due to exposure to fire. In one instance, refitting showed one discolored flake

[1] This hominin is currently known as *Australopithecus africanus*, with an estimated age of ca. 3 Ma.

[2] The earliest specimens of *H. erectus* (*sensu lato*) are dated to about 1.9 Ma in Africa. This is recognized as the first hominin to leave Africa and spread throughout the Old World, where it persisted until about 0.5 Ma (O'Connell et al. 1999 and references therein; Ungar et al. 2006 and references therein).

while other blanks from the same core were unchanged (Clark and Harris 1985). Artifacts modified on basalt and chert were found in some of these "fireplaces". The TL response of some of these pieces demonstrated that they were heated contemporaneously with the "fireplace" (Rowlett 2000).

It is interesting to note that the spatial distribution of the archaeological material demonstrates that the highest concentration of lithics and bones is in the vicinity of the patches of burned sediments (Clark and Harris 1985).

Excavations at the second locality, FxJj 20 Main, have revealed two oxidized features; one of these was proposed to represent the remnants of an ancient fireplace. These features contained fully oxidized sediments to a depth of at least 5 cm (Bellomo 1994a). Analyzed samples originating from these features exhibited magnetic susceptibility typical of fireplaces, suggesting that "individual campfire sites were used on more than three or four occasions, or that the fires at those sites were maintained for a duration of at least a few days" (Bellomo 1994a:17). In addition, thermal alteration was observed on three of the 335 stone artifacts recovered from the vicinity of the oxidized features at FxJj 20 Main (Bellomo 1994b). These were two chert artifacts with reddish and yellowish spot discolorations, and one basalt artifact that appears to contain evidence of potlidding (Bellomo 1994b).

Several methods of spatial analysis (e.g., nearest-neighbor and local density analysis) were employed to determine whether the observed spatial patterning resulted from hominin activities or post-depositional processes. The combined data suggest that "...the campfires provided a central focus of activities, including the production and maintenance of stone tools and the consumption of food" (Bellomo 1994b:194).

Other evidence for fire derives from the site of *Chesowanja*, located near the east shore of Lake Baringo at the foot of the Laikipia Escarpment in the Kenya Rift (Gowlett et al. 1981; Isaac 1982). Evidence for fire was found in the locality of GnJi 1/6E, which underlies a basalt flow dated to ca. 1.4 Ma (Gowlett et al. 1981). The evidence consists of 40 pieces of burned clay, ranging in size from small flecks to 5–7 cm lumps that were exclusively found intermingled with Oldowan stone tools and animal bones (Gowlett et al. 1981). Samples of the burned clay were examined by magnetic susceptibility, which concluded that these are the result of heating at 400°C, a normal temperature for open camp fires. Analysis of the burned clay concluded that "...the Chesowanja clay was burned by a small, controlled fire" (Gowlett et al. 1981:128).

Clark and Harris (1985) describe some 51 reddish-brown clasts of clay from this site, the largest pieces of which were concentrated in an area of 3 m^2 together with a high proportion of the cores and cobbles. Provided that this is a cluster in situ, the association of burned clay and cobbles may suggest that these are the remnants of a hearth. A sedimentological study indicated that the silty clay sediments at the site were deposited in a low-energy fluvial environment. However, small pockets of coarse sandy sediments, interstratified and interfingered with the silty clays, suggest that high rainfall and sheet wash have rearranged and concentrated the archaeological material in a small runnel (Clark and Harris 1985). In addition to this spatial configuration, the burned clay was found linearly distributed within a possible runnel feature. However, no significant size sorting of the stone artifacts was found (Clark and Harris 1985).

The site of *Gadeb*, situated near the high western edge of the Southeast Plateau of Ethiopia, displays more possible evidence of burning. The site is embedded in a series of lacustrine and fluviolacustrine sediments of Plio-Pleistocene age that were dated to the range of 2.7–0.7 Ma; the archaeological occurrences date from 1.5 to 0.7 Ma (Clark and Harris 1985). Evidence of fire was found at the Acheulian site of *Gadeb* 8E in the form of weathered angular fragments of tuff with differential dark gray and red discoloration. Although these presumably burned rocks occurred singly, a group of four such fragments was found distributed in 1 m^2 (Clark and Harris 1985). Ten of these rocks were subjected to paleomagnetic analysis and all were found to have a magnetization of thermal origin (Barbetti 1986). It is possible that these rocks retain a thermal magnetization from the time of original formation. However, the directions of magnetization of the stones as recorded in the field were not random, but rather point to a uniform direction. The analyses thus concluded: "The palaeomagnetic results from Gadeb put the weight of evidence marginally in favour of fire" (Barbetti 1986:778).

Excavations in the *Middle Awash*, along the Awash River Valley of Ethiopia, yielded several archaeological occurrences dated to 2.0–0.5 Ma, some with evidence for fire. In the vicinity of the Oldowan site of BOD-A4 and the Acheulian site of HAR-A3, clay samples were collected from cone-shaped reddish areas ranging from 40 to 80 cm in diameter (Clark and Harris 1985). Since the clay was generally more resistant to erosion than the surrounding sediments, these areas were found in the form of small, low mounds some 20–30 cm high (Clark and Harris 1985). On the basis of paleomagnetic analysis, the clay samples from these two sites were interpreted as having been baked at temperatures of 600°C or more (Barbetti 1986). It is, however, uncertain whether these features are the result of fireplaces. It is interesting to note that although the clay patches were found in association with lithic artifacts and bones, none were found within the burned sediments. This is one of the reasons for the prevalent interpretation that these burned sediments are the result of burning tree stumps and that the burned clay is termite earth that was on the stump at the time it was burned (Clark et al. 1984; Clark and Harris 1985).

A hearthlike feature was observed during excavations at the Acheulian site of *Olorgesailie* in Kenya. The "hearth" was a depression filled with lithics and bones, although no charcoal was detected in it (Isaac 1977). Microscopic fragments of charcoal were observed during the search for pollen grains. However, it was uncertain whether these are the result of human activity or wild bush fires (Isaac 1977). Isaac concluded that the hearthlike feature is not conclusive evidence and that "...the Olorgesailie sites lack positive traces of the use of fire, in the form of charcoal, visibly burned bone, or obvious hearth structures" (Isaac 1977:93).

In the more recent investigations of the Olorgesailie Basin, samples of such "reddened zones", which were found to occur at several levels within the upper part of the Olorgesailie Formation, have been collected and analyzed. The analysis has focused on in situ reddened and partially fused areas and sampled two types of sediments; reddened samples, slightly melted or not melted, and melted rocks that have undergone nearly total fusion (Melson and Potts 2002). The various mineralogical and chemical analyses of these samples (dated by Ar/Ar to ca. 0.6–0.5 Ma) indicate that they are products of near-surface underground combustion; the presence of the melted rocks is suggestive of a period of extreme drought, in which extremely dry conditions led to subsurface fires, combustion metamorphism, and the formation of these burned features (Melson and Potts 2002).

Various burned plant materials (e.g., charred logs, charcoal, carbonized grass stems and plants) were recovered from the waterlogged Acheulian occupation at *Kalambo Falls* in northern Zambia (Clark 1969, 2001). In addition, rare fire-fractured quartzite items were found (Clark and Harris 1985).

The use of fire has been linked with the early occupations of South Africa (Beaumont and Vogel 2006:226), where several cave sites have demonstrated early evidence of fire. Excavations at the South African cave site of *Swartkrans* unearthed a sequence of Early Stone Age occupations. An assemblage of blackened bones was recovered from the Acheulian horizon of Member 3 (Brain and Sillen 1988), dated to 1.9–1.65 Ma (Delson 1988). Comparison of these bones with burned bones from experimental burning suggested that the former are the result of intentional burning to various degrees; some of the bones were slightly heated to a temperature below 300°C, some to 300–400°C, others to 400–500°C, and most to a temperature above 500°C (Brain and Sillen 1988). Renewed analyses of the bones with the use of the Electron Spin Resonance (ESR) technique have confirmed these observations, suggesting that the hominins of Swartkrans used fire, and that the burning of the bones is possibly the result of cooking/roasting of meat (Skinner et al. 2004). The evidence from Swartkrans represents one of the best early contexts for investigating the emergence of fire use. Furthermore, recent analyses of the faunal assemblages have indicated that the Swartkrans hominins were not only fire users but also adept procurers of ungulate carcasses who were able to gain access to and exploit "...the most nutritious components of those resources before potential competitors" (Pickering et al. 2008:42).

Similarly to Swartkrans, evidence from *Wonderwerk Cave* consists of "...hundreds of charred-calcined large-mammal bone fragments" (Beaumont and Vogel 2006:222). At Wonderwerk these were found embedded in an extensive sheet of wood ash dated to ca. 1.1 Ma (Beaumont and Vogel 2006).

At the South African site of *Cave of Hearths*, burned deposits were observed from the base of the archaeological sequence (Early Stone Age) through five stages of the Middle Stone Age and up to the occupational horizons of the Late Stone Age. The exposure of the "basal hearth" in the third Acheulian horizon uncovered a thick (1.3 m) ash deposit, transformed into breccia. Fragments of bones were found within the ashy sediments and their presence was interpreted as follows: "They apparently chewed or broke animal bones into small pieces and threw them into the fire..." (Mason 1969:159). In addition, two handaxes from this area were reported to be fire-pitted (Oakley 1954). Following an analysis of some samples from the basal hearth, Oakley (1954) concluded that unlike other hearths in the upper parts of the cave's sequence, the basal hearth was devoid of free carbon, thus suggesting that the sediments are not wood-ash. Rather, the sediments consist of bat guano which was either used as fuel for hearths (Oakley 1954) or struck by lightning and thus turned into ash (Latham and Herries 2004). Analyses of these sediments using contemporary methods may provide conclusive evidence for the question of fire use at the Cave of Hearths (Latham and Herries 2004).

1.1.2 The Levant

Several Levantine Middle Paleolithic sites have produced pioneering sedimentological studies that enable the identification of burned sediments and thus of ashes and hearths (e.g., Schiegl et al. 1994, 1996; Albert et al. 1999, 2000, 2003; Weiner et al. 2002). Such analyses are rarely available from Lower Paleolithic occupations (but see Karkanas et al. 2007), and thus the early Levantine evidence is based mostly on the presence of burned lithics and bones.

At the 1.4 Ma old Acheulian site of *'Ubeidiya* in Israel, evidence for fire is derived from several flint implements. Thirty-one burned flint artifacts originating from 14 different archaeological horizons were observed (Bar-Yosef and Goren-Inbar 1993), either within a defined living floor or as sporadic scatters of artifacts within the deposits. In some cases the burned items were fragments of a single piece that was shattered by fire and found in a limited area (Bar-Yosef and Goren-Inbar 1993). However: "Such scanty evidence does not permit further speculation on the possibility that fire was used by the 'Ubeidiya hominids" (Bar-Yosef and Goren-Inbar 1993:191).

Excavations of the Acheulian archaeological horizons at *Latamne* in northern Syria unearthed concentrations of limestone blocks and angular rubble of flint and limestone. Examination of the depositional environment concluded that it would have been impossible for these to have been deposited naturally at the site (Clark 1966). Influenced by Stekelis's report of burned bones from Gesher Benot Ya'aqov (Stekelis 1960), Clark suggested that "the rubble concentration at Latamne could be explained as having been used in the construction of stone 'ovens' for cooking meat and vegetable foods…" (Clark 1966:219). Some of the limestone blocks exhibited fractures, reddening and discoloration, features similar to those resulting from exposure to fire. One such limestone item was examined by Oakley and tests for thermal alteration yielded negative results (Clark 1968). Additional evidence can be found in some flint artifacts which display thermal damage, such as potlid fractures and occasional reddening, suggestive of thermal alteration. However, as suggested by Clark, these features might be the outcome of frost action (Clark 1966).

The site of *Bizat Ruhama*, situated on the eastern margin of Israel's southern coastal plain, is dated to ca. 1.0 Ma (Zaidner et al. 2003). Although faunal remains were retrieved from various areas of the site, in only one area did the bones differ in color and preservation by being fragile and whitish. This feature, in addition to the presence of scant charcoal fragments in that area, suggested that the bones are burned (Ronen et al. 1998).

Excavations at *Tabun Cave* in Mount Carmel (Israel) yielded an extensive cultural sequence of Lower and Middle Paleolithic occupations. Within the lower part of the sequence, evidence for the use of fire can be found in the Acheulo-Yabrudian horizons (Ea-d). TL dates on burned flints suggest a date for these levels of 0.35–0.3 Ma (Mercier et al. 1995), although a somewhat older date, a combined ESR/U-series age of ca. 0.39 Ma, has also been proposed (Rink et al. 2004). The TL dating provides the only reference for the presence of burned lithics in the Acheulian assemblages of Tabun. However, during excavation of layer E, faintly colored but well-defined hearths were observed by Garrod throughout the layer (Garrod and Bate 1937). The hearths of layer E differed from the surrounding sediments and were darker brown or yellow; some of them appearing to be more intensive: "Scattered all over E were patches, more or less extensive, of white crumbly earth containing badly calcined flint, which presumably mark the place of particularly intensive fires" (Garrod and Bate 1937:66).

Garrod observed additional evidence of the use of fire during this time range at the rock shelter of *Abri Zumoffen* in Lebanon. The evidence consisted of hearths in which flint was sparse and relatively large bones occurred (Garrod and Kirkbride 1961). These "…intact hearths" are also reported by Copeland (1983:76, 2000:98).

A hearth associated with Yabrudian lithic artifacts and faunal remains was also reported from *Bezez Cave* in Lebanon (Kirkbride 1983).The hearth was some 10 cm thick and several "…heat-fractured artifacts were found in its vicinity" (Kirkbride 1983:31). The presence of additional possible hearths "…may be inferred from burned flint and bone, and charcoal scraps…" (Kirkbride 1983:31).

Acheulo-Yabrudian deposits were also recently discovered at the Mount Carmel cave of *Misliya* in Israel (Weinstein-Evron et al. 2003). Burned flint artifacts retrieved from these deposits are being subjected to TL analyses (Y. Zaidner 2007, personal communication).

Other Acheulo-Yabrudian deposits discovered at *Qesem Cave* in Israel have recently been dated by uranium isotopic series to the range of ca. 0.38–0.20 Ma (Barkai et al. 2003). Evidence of fire includes burned bones, lithics, and sediments, and occurs throughout the 7.5 m deposit (Stiner et al. 2004). In the lower part of the sequence "only discrete lenses of burnt remains are observed" (Karkanas et al. 2007:10) while in the upper part of the sequence evidence for fire use consists of thick ash-rich deposits, associated with large amounts of burned bone fragments and lumps of heated soil (Karkanas et al. 2007:10). The evidence from Qesem Cave joins that from the sites discussed above, which suggests a consistent use of fire towards the end of the Levantine Lower Paleolithic.

The presence of burned items or features is not noted in the publications of other Levantine Lower Paleolithic sites. However, a flake displaying potlidding is illustrated in Neuville's report of the Tayacian assemblage of E3 at *Umm Qatafa* in the Judean Desert of Israel (Neuville 1951: Fig. 13). Burned flints, particularly small ones, are present at the site of *Revadim* (Israel) (personal observation), dated to ca. 0.3 Ma (Marder et al. 1998).

1.1.3 Asia

The Asian continent has yielded what was long considered the earliest evidence for the use of fire at the Chinese site of Zhoukoudian. However, the evidence from *Zhoukoudian* is currently controversial due to recent mineralogical analyses of the site's sediments. The site displays a series of cavities located on the Dragon Hill some 50 km southwest of Beijing. Major significance is attributed to Locality 1, in which the remains of "Peking Man" and his related assemblages were supposedly embedded in layers of dark ashes dating from ca. 0.6 to 0.3 Ma (Goldberg et al. 2001). These early indications of fire, particularly the 4–6 m accumulation of "ashes" in Layer 4 and the "hearth" of Layer 10, have been extensively discussed and were long interpreted as representing the remains of hearths constructed and used by man (e.g., Breuil 1932; Stewart 1956; Oakley 1956, 1961; see also Goldberg et al. 2001:518–520 and references therein).

However, recent mineralogical analyses of the sediments from Locality 1 suggest that no thick ashy accumulations or even ash remnants (i.e., siliceous aggregates) are present (Weiner et al. 1998; Goldberg et al. 2001). According to these analyses (Goldberg et al. 2001), the dark "hearth" in Layer 10 is a deposit composed of finely laminated unburned organic material interbedded with silts, and the bulk of Layer 4 is bedded to laminated silts of loessial origin that were washed into the depression. During the renewed analysis of the "ashy" sediments, burned bones, in association with lithic artifacts, were found exclusively in the upper unit of Layer 10, thus suggesting that "this association of the burned bones and artifacts constitutes possible, but not conclusive evidence for fire use by humans at Locality 1" (Goldberg et al. 2001:520). These burned bones were either black or turquoise-colored and were interpreted as "fossil bones that were somehow burned by natural processes" (Weiner et al. 1998:252). Thermoluminescence (TL) analysis of fire-cracked hammerstones and burned hackberry seeds (*Celtis barbouri*) from Locality 1 suggested that these are "clearly heat crazed and carbonized respectively, so obviously they have been burnt somewhere at some time" (Rowlett 2000:207).

Until recently, the evidence from Zhoukoudian was considered the best example of the use of fire by early man. Following the mineralogical analysis, this evidence is now inconclusive. However, it is important to note that sediments from Zhoukoudian were dated by the TL method. In Layer 10, in which burned items were found, the TL glow was lower than in sediments originating from layers in which no burned items were recorded (Rowlett 2000).

Charred wood remains were found at the site of *Trinil* in Java. Potassium–argon dating of the site suggested an age of 0.8–0.5 Ma and later a reading of 1.2 Ma (James 1989 and references therein). Oakley (1956) suggested that these charred wood remains were the result of natural fires: "Volcanic activity in this region probably caused forest fires from time to time during the accumulation of these deposits" (Oakley 1956:40).

The site of *Xihoudu* in China yielded a large faunal assemblage along with some 30 lithic artifacts. Some of the bones were black, gray and grayish-green and are considered as burned on the basis of laboratory analysis (James 1989). The faunal remains are considered to be some 1.0 Ma old while paleomagnetic readings suggested an earlier date of 1.8 Ma (James 1989).

A *Homo erectus* cranium together with 20 stone artifacts was found at the Chinese site of *Gongwangling* (James 1989). Magnetochronological studies dated the site to ca. 1.2 Ma (Hyodo et al. 2002) and the presence of several charcoal flecks at the site is suggestive of burning (James 1989).

Excavations at the Chinese site of *Yuanmou* unearthed two *Homo erectus* incisors and faunal and lithic material (James 1989). Evidence for the use of fire is represented by the dark color of two of the mammal bones and the considerable amount of charcoal found at the site (James 1989). Recent magnetochronological studies date these remains to 0.7 Ma (Hyodo et al. 2002).

Pope (1983) mentions evidence of fire from the site of *Lantian* (Chenjiawo) in China at ca. 0.78 Ma, although the type of evidence that is preserved is not reported.

1.1.4 Europe

The European evidence also incorporates several sites in which the evidence of fire is fragmentary and controversial. The site of *St. Estéve-Janson* (Escale Cave) in France's southern Durance Valley yielded Middle Pleistocene faunal remains and a few limestone flakes alongside hearths, fire-cracked rocks, ash, and charcoal (James 1989). The five hearths were represented by reddened areas, a meter in diameter. Paleomagnetic analysis on samples from these reddened sediments suggested that these are burned (James 1989). However, the only cultural material found in association with these "hearths" is several limestone flakes. It is thus unclear whether these features are actually the result of human activity.

Howell reported on some possibly burned flints from the site of *Montières* in France. The site, embedded in the Pleistocene sediments of the Somme Terrace, revealed several worked stones. Some of these had a "porcelainized aspect as if subjected to fire" (Howell 1966:91).

Possible use of fire at the English sites of *Swanscombe*, *Hoxne*, and *Marks Tey* was implied by changes in pollen frequencies and the presence of charcoal fragments in the deposits. At Swanscombe, dated to ca. 0.3 Ma, lumps of carbonized vegetable material were found. These were described by Oakley (Oakley 1956) as charcoal resulting from "fires burnt on the banks of the river by Acheulian hunters" (Oakley 1956:41). Reddened and crazed flints were initially determined by Oakley as burned, but after later studies Oakley concluded that there are no burned flints or bones at the site (James 1989). Palynological studies of the Hoxne and Marks Tey deposits indicated a decrease in arboreal pollen and an increase in grasses in the Acheulian Layer E. Arguing that climatic changes could not account for these shifts in vegetation, and based on a single piece of charcoal found at Hoxne, it was suggested that hominins induced forest fires for hunting purposes (James 1989).

A recurrent difficulty when discussing early European sites is the classification of early "pebble industries" as being of anthropogenic or natural origin (e.g., Roebroeks and van Kolfschoten 1995). This is the case at the site of *Blassac-les-Battants* in France, where faunal remains dated to 1.4–1.2 Ma were found in association with lithic objects (Raynal et al. 1995). Although some of these crystalline rock items exhibit obvious thermal fractures, the entire "assemblage" is most likely naturally fractured (Raynal et al. 1995).

Similarly, the site of *Přezletice* in the Czech Republic revealed Pleistocene deposits dated paleomagnetically to the range of 0.89–0.59 Ma (Valoch 1995). Charcoal remains, burned bones, burned stones, and the remains of a fireplace are reported from the site (Valoch 1995). However, the lithic artifacts of Přezletice display vague flaking properties that in the opinion of some scholars do not show convincing traces of human intervention (Roebroeks and van Kolfschoten 1995).

A related problem occurs at the site of *Šandalja Cave I* in Croatia, where charcoal and burned bones were documented in Early Pleistocene breccia (Valoch 1995). Only two lithic items were found in association with these finds, an unmodified pebble and a flint chopper, which do not allow an unambiguous interpretation of the site (Valoch 1995). Valoch (1995) also reports on burned bones from *Stránská Skála I* in the Czech Republic. The bones were found within the early Middle Pleistocene archaeological horizons, and chemical analyses suggest that they were indeed burned to 200–500°C (Valoch 1995).

Indisputable evidence of early use of fire in Europe emerges from the renewed archaeological investigations at *Beeches Pit* in Suffolk, England. These have revealed varied evidence of burning for which different dating methods all suggested an age of ca. 0.4 Ma (Gowlett 2006; Gowlett et al. 2005). The evidence includes reddened and crazed burned flints, which on the basis of the TL method appear to have been exposed to temperatures above 400°C (Gowlett 2006). Exposure to temperatures of 600–800°C is suggested for the numerous charred and calcined bones from the site; such intensity of burning may be the outcome of the use of the bones as fuel (Preece et al. 2006). Burning is also observed in shells and charcoal (Preece et al. 2006). In addition, the sharply delimited dark fill with reddened sediments underneath and/or at the margins most likely represents the remnants of hearths (Gowlett 2006; Preece et al. 2006, 2007). The spatial disposition of artifacts at Beeches Pit, particularly the refitting series, is associated with these burned areas (Preece et al. 2006, 2007).

Excavations at the site of *Menez-Dregan* (Brittany, France) have uncovered a hearth in the form of a deep concentration of charcoal and burned bones. ESR dating of burned quartz suggested a date of 0.46 Ma, which is supported by stratigraphic and micromorphological analyses (Monnier et al. 1994; Patel 1995; Geigel et al. 2004). The fire at Menez-Dregan is presumed to have functioned at low to moderate temperatures, enabling the preservation and analysis of ancient DNA in fossil bones (Geigel 2002; Geigel et al. 2004).

At the site of *Schöningen* in Germany, which is dated to ca. 0.4 Ma, wooden spears that were apparently fire-hardened at the tips (Thieme 1997), as well as burned flints and hearths (H. Thieme 2004, personal communication), were found.

Substantial evidence of fire use is reported from the site of *Vértesszőllős*, located along the fault line of the western Gerecse Mountains in Hungary. The site, described as a "*Sinanthropus*" camp site, revealed the remains of two hominins that were given the name *Homo erectus* seu *sapiens palaeohungaricus* (Thoma 1990). The ancient footprints of these hominins were found on the living floor along with fireplaces, lithic artifacts, and faunal and botanical remains (Kretzoi and Dobosi 1990). Th/U analyses of the travertine sediments of Vértesszőllős suggested an age of 0.35 Ma, and ESR dating of the travertine yielded a similar age of 0.33 Ma (Pécsi 1990). The documented fireplaces all contained fragmentary burned bones but no charcoal. Interestingly, the burned bones were laid in a radial fashion around the center of the fireplaces (Vertes and Dobosi 1990). Observations during field work describe the fireplaces as follows: "The fireplace itself has a slightly domed appearance. The highest point is in the middle. When its profile was examined it turned out that the fireplace was of the same thickness overall: It was the base of the feature which had the protruding center. It appears that the fire itself was built in a pyre-like form or was covered by bones placed neatly side-by-side. Several other similar fireplaces were also discovered. They usually measure 30 × 40 cm. The largest discovered to date is 35 × 45 cm in size. Their outlines are irregular and they are located at unequal distances from each other … the thickness of the fireplaces ranges from 3 to 5 cm" (Vertes and Dobosi 1990:520). The presence of burned and fragmentary bones and the lack of charcoal in these fireplaces led to the assumption that bones were used as fuel. An experimental fireplace, constructed with bone fragments, was proven to be particularly advantageous in a wet environment such as that of Vértesszőllős (Vertes and Dobosi 1990). According to James (1989), these fireplaces might merely represent mineral staining from groundwater. However, given the radial arrangement of the bones and the fact that both burned and non-burned bones are present at the site, the original interpretation of these features seems more plausible.

A similar interpretation was suggested for the burned bones from *La Cotte de St. Brelade* on Jersey Island in the English Channel. The site was dated to the range of 0.38–0.2 Ma (Huxtable 1986) and yielded varied evidence for the use of fire. Indications of burning are reported from all occupation layers and comprise charcoal, burned bones, burned flint artifacts, and burned granite (Callow et al. 1986). Several small patches of fire-reddened earth were observed and identified as remnants of hearths; the high frequencies of burned bones and the predominance of these over wood charcoal suggested that bones were used as fuel (Callow et al. 1986).

Evidence for fire is recorded from the Spanish site of *Torralba*, dated to ca. 0.35–0.3 Ma. Situated midway between the cities of Madrid and Zaragoza, the site lies on the Rio Ambrona-Masegar valley cut by the Ambrona River. Excavations at *Torralba* exposed an area of over 30 m^2 and uncovered the semiarticulated remains of the left side of a large elephant, without a pelvis but with some vertebrae, tusks, and a complete mandible (Howell 1966). Only four retouched flakes were found in association with the elephant. An area to the southeast yielded other remains of the same individual in addition to bovid remains. In that area, several patches of charcoal were observed (Howell 1966), suggesting that the processing of the elephant's meat involved the use of fire. In total, some 232 fragments of charcoal are known from Torralba, in addition to hundreds of near-microscopic fragments (Freeman 1975). The unique conditions of preservation enabled the preservation not only of charcoal but of some 76 wooden fragments and 31 casts of decayed large wooden objects, some with deliberate cultural alteration (Freeman 1975). One of these wooden objects is a trapezoidal block (12 × 9.3 × 3.7 cm) with a darkened coloration suggestive of burning (Howell 1966). It has been suggested that the wood was used as fuel for the maintenance of fireplaces at Torralba. This is suggested by the fact that the abundance of wood fragments is directly correlated with the abundance of charcoal. Furthermore, there is a tendency for wood and charcoal to cluster together spatially in discrete clumps (Freeman 1975).

Excavations at *Bolomor Cave* (Valencia, Spain) have uncovered what is currently the most ancient known evidence for the use of fire in the Iberian Peninsula (Fernández Peris 2003). The lowest level of the stratigraphic sequence at Bolomor (XVII) is estimated at 0.35 Ma; hearths were uncovered in level XI, where they appear as distinct marks of dark soil associated with fragments of burned bones (H. Fluck 2007, personal communication).

In Cantabria, the early Middle Pleistocene site of *San Quirce* revealed an abundant lithic industry without faunal remains. The lithic material was found "clustered spatially and in association with ash and a possible hearth" (Raposo and Santonja 1995:10). A more plausible hearth is reported from the Middle Pleistocene site of *Solana del Zamborino* near Granada. The hearth is defined by "a circle of five quartzite pebbles, with an impressive amount of charcoal and ash in the middle" (Raposo and Santonja 1995:19).

Located along the Mediterranean shore in the city of Nice in France, excavations at the site of *Terra Amata* exposed what is acknowledged as the most ancient example of built structures with interior hearths. TL dates suggest a date of 0.25–0.2 Ma, while correlation of the geological sequence with Isotopic Stage 9 suggests a date of 0.33 Ma (Villa 1983); additional TL dates suggested an age of 0.38 Ma (Scarre 1998). The hearths were found in the centers of huts, of which only the postholes remained, exhibiting areas of reddened sand about 30 cm wide with traces of charcoal and reddened pebbles. In some cases, a small pile of pebbles was found near the hearth, supposedly to protect the fire from drafts (Villa 1983). Concentrations of charcoal (mostly identified as *Pinus sylvestris*), as well as burned flints, burned mussel shells (Villa 1983), and burned bones (Villa 2001), were also observed.

A similar association between hearths and dwelling structures is reported from the site of *Bilzingsleben* in Germany, dated to 0.3 Ma, where the foundations of three simple dwelling structures with hearths in front of them were observed (Mania 1995). Charcoal, burned stones, and burned bones are also reported (Villa 2001). Similarly, at *Azokh (Azych) Cave* in the Caucasus, two Acheulian layers were exposed within the Middle Pleistocene deposits. Four hearths were found, one of them located inside a limestone dwelling feature (Ljubin and Bosinski 1995).

1.1.5 Summary

Review of the reported evidence demonstrates that great efforts have been made to identify archaeological evidence of fire, and hence to identify the initial stages of human control over fire. The various indications suggest that this stage in our evolution occurred in Africa some 1.5 Ma. *Homo erectus* (*sensu lato*) was most likely our first ancestor to overcome the fear of fire and to "domesticate" it to his needs. It was at this time that *Homo erectus* started to explore new territories and the human migration out of Africa was initiated. The chronological and geographical distribution of the earliest evidence of fire (Fig. 1.1) suggests that fire could have been a stimulating tool during this stage of human dispersal (see Chapter 4 for further discussion).The reviewed evidence of early use of fire is based on highly varied criteria for the identification of fire. Likewise, a wide diversity of phenomena is used to determine the early use of fire. This diversity is further emphasized in the following section, which attempts to explore the varied methods and techniques used to identify the early archaeological evidence of fire use.

1.2 Identifying Fire: Methods and Techniques

The review of the early evidence for the use of fire has revealed a great variety of indications. Perhaps the most striking of these is the diversity of evidence used to determine man's use of fire: burned items (e.g., lithics, bones, wood, and shells), burned sediments, ashes, and charcoal. The following discussion incorporates data from a larger time span than that of the early evidence. It demonstrates that both the criteria by which controlled fire is recognized and the means by which the evidence is analyzed are diverse.

It is important to note that in using the term "*controlled*" we refer to anthropogenic fires, modified and maintained by

Fig. 1.1 Chronological and geographical distribution of the major occurrences of early fire mentioned in the text

humans. The term "*controlled fire*" is thus used here to describe small-scale "*domestic*" fires, unlike the ecological term that denotes large-scale controlled burning designed for clearing of vegetation.

1.2.1 Burned Sediments

At some of the sites at which early evidence for the use of fire is documented, the identification of burned sediments occurred during field work (e.g., Beeches Pit, Koobi Fora), and areas of reddened soils are often interpreted as indications of hearths. Experimental studies have shown that fireplaces can result in discoloration of the sediments to dull yellow, red, or black on the surface directly below the fire (Bellomo and Harris 1990). It has been suggested, however, that these alterations may disappear from the soil after a long period of weathering and leaching (Bellomo and Harris 1990).

Other experimental works have demonstrated that fire does not necessarily result in discoloration of sediments (Canti and Linford 2000). These studies confirmed that sediments beneath the experimental hearths remain below 500°C, so that reddening of the soil rarely takes place (Canti and Linford 2000). In addition, the process of iron oxide transformations, which causes soil reddening, varies among different sediments, possibly due to organic matter content, chemical variations in sediments, soil moisture, or the fuel used (e.g., Canti and Linford 2000; Linford and Canti 2001; Leesch et al. 2005 and references therein). It thus appears that discoloration of sediments is not a sufficient measure (when used independently of other lines of evidence) for the identification of burned soils. As discussed above, the discoloration observed at Locality 1 at Zhoukoudian, long considered an indication of fire, was recently rejected on the basis of sedimentological analyses (Weiner et al. 1998; Goldberg et al. 2001).

Sedimentological analyses have proved to be reliable in determining the presence of burned elements within archaeological sediments. Such studies have focused mainly on Middle Paleolithic cave deposits that display visible ash accumulations. A major mineralogical component of the ashy sediments was found to be siliceous aggregates, which are present in wood and similar to those present in the residue of fresh wood ash (e.g., Schiegl et al. 1994, 1996; Elbaum et al. 2003). Sedimentological analyses have also enabled the identification of wood ash components based on the phytolith composition. These microscopic plant silica bodies can determine the presence of ashes and further establish whether wood, bark, or grasses were used as fuel for the fireplaces (e.g., Albert et al. 1999, 2000). At Middle Paleolithic cave sites in France, sedimentological studies have enabled the identification of the use of lichen as fuel (Rigaud et al. 1995) as well as lignite collected from natural outcrops within 7–15 km of the sites (Théry et al. 1996).

Despite the clear contribution of the different sedimentological analyses, they necessitated sampling from suspected in situ burned features, and such features are only preserved in favorable conditions. This is demonstrated by the preponderance of studies from cave sites. Yet, following the archaeological identification of burned features within the sediments, the use of such analyses can contribute valuable supportive evidence to the observed burning.

As previously discussed within the review of the African sites, burned sediments can also be subjected to magnetic analyses (e.g., Koobi Fora, Chesowanja, Gadeb, Middle Awash) in order to determine whether they are the result of burning (Bonhomme and Stanley 1985; Barbetti 1986; Jordanova et al. 2001). Through burning, hematite (Fe_2O_4) in the soil becomes reduced to the more magnetic magnetite (Fe_3O_4), and where these minerals have been heated above their Curie temperature and then cooled, they acquire remanent magnetism (measurable in the lab). In addition, measurement (in the field or in the lab) of the magnetic susceptibility of burned sediments through examination of enhanced susceptibility of magnetic magnetite (Fe_3O_4) over Fe_2O_3 can suggest heating of the sediments.

Other means of determining burning of sediments are Thermoluminescence (TL) and Electron Spin Resonance (ESR) analyses. Both methods can provide a measure of the number of electrons trapped within radiation-induced defects in solids. Since the concentrations within such defects increase with time as a result of exposure to natural radiation in the environment, these methods are used for dating (e.g., Mercier et al. 1995; Valladas et al. 1998). The radiation-induced defects (the amount of light given off during heating in the case of TL, or the amplitude of a derivative line of an ESR spectrum) are destroyed (annealed) during heating, and thus their measurement enables an estimation of whether heating occurred (Bischoff et al. 1984). These methods can be used to determine burning in ashy sediments (e.g., Bischoff et al. 1984), as well as in burned stones (e.g., Hedgcock et al. 1988; Alperson-Afil et al. 2007), burned bones (e.g., Skinner et al. 2004), and botanical remains (e.g., Hillman et al. 1983).

1.2.2 Burned Bones

The presence of burned bones in archaeological occupations is suggestive of fire use at the site. Burning of the bones can be the result of their use as fuel for hearths (e.g., Vértesszőllős: Vertes and Dobosi 1990; see also Théry-Parisot 2001; Villa et al. 2002, 2004), of cooking/roasting of meat (e.g., Swartkrans Cave: Brain and Sillen 1988; Skinner et al. 2004), or of their random proximity to hearths (e.g., Champréveyres and Monruz: Leesch et al. 2005), including bones embedded in the subsurface beneath the fire (Bennet 1999). Identification of burned bones can be based on the visible discoloration of bone, changes in bone mineral and matrix, and changes in the mechanical properties of bone (Shipman et al. 1984; Nicholson 1993; Villa et al. 2004). Changes in mechanical properties can increase fragmentation of bones, so that the bulk of burned bones is likely to be of smaller size than that of the unburned assemblage (Stiner et al. 1995; Villa et al. 2002 and references therein).

Recently, Hanson and Cain (2007) have noted that when macroscopic analyses are used alone, certain modifications of bones caused by physical and biological post-depositional processes may be confused with burning. Hanson and Cain have consequently established a new methodological approach, which suggests that histological evidence of burning is preserved through fossilization and diagenetic processes, so that analysis of the microscopic internal structure of bones is a valid method for distinguishing burned from unburned bone (Hanson and Cain 2007).

Burned bones are thus an efficient indication of the use of fire. However, the preservation of bones varies between sites, as do depositional environments. While blackened bones at one site can indicate fire, at another they may be the result of the depositional environment (e.g., Shahack-Gross et al. 1997). Thus, our ability to identify early fire by using fossil bones is limited by their state of preservation.

1.2.3 Burned Stones

Exposure to fire may change the mechanical properties of lithics. Experimental studies have been carried out mostly on flint and chert and demonstrate that exposure to high temperatures (~350–500°C) causes various alterations such as discoloration, potlid fractures, crazing, and fragmentation (e.g., Purdy and Brooks 1971; Purdy 1975, 1982). In addition, an experimental study restricted to burning of flint was recently carried out (Sergant et al. 2006) and suggested that only direct contact with fire results in the formation of distinctive heat damage on flint. In these experiments, open hearths were constructed on a flat, sandy surface without vegetation. The hearths consisted of a basal layer of small pine twigs covered by pine branches that were laid in a radial formation at a diameter of ca. 70 cm. An unbroken line of flint artifacts was placed from the surface hearths to the surrounding area and temperatures of the artifacts were taken during heating. This experiment has demonstrated that "only those artifacts which were in direct contact with the fire were heated to a temperature above 300°C and showed heat damage. Artifacts lying outside the hearth, even those immediately bordering the hearth, were not affected by the heat at all" (Sergant et al. 2006:1001).

For sites in which this raw material occurs, the presence of fire can be inferred through the identification of burned flints. This method is particularly advantageous as it draws on features of burning damage that are easily identified by the naked eye. Furthermore, not only does burned flint attest to the presence of fire, it also indicates the temperatures of the fire (i.e., above 300°C) and the fact that the burned items must have been exposed on the surface in order to have direct contact with the fire. Accordingly, given that the burned flints are found in situ, their spatial setting can be evidence of the location of the fire.

The various methods and techniques in use further demonstrate the great scientific efforts made to identify controlled use of fire in archaeological sites. Clearly, the identification of burned materials indicates the presence of fire. However, when attempting to infer human-controlled fire, the mere presence of burned items or features is not sufficient. In addition, as varied as the methods are, certain lines of evidence are bound to be absent in sites in which post-depositional processes have concealed the evidence: "taphonomic problems related to the preservation of ash (Schiegl et al. 1996), charcoal (Cohen-Ofri et al. 2006), and other indications of fire use (e.g., burnt bone; Shahack-Gross et al. 1997) hinder the discovery and recognition of burnt remains" (Karkanas et al. 2007:198). These taphonomic complexities constrain the attempts to identify early human use of fire and actually suggest that of the various possible indications, burned lithic artifacts are the most durable component of the archaeological record that is available for such analyses.

The evidence for the early use of fire at the Acheulian site of Gesher Benot Ya'aqov includes burned flint artifacts, charcoal fragments, burned wood, fruits, and grains. In addition, small burned flint artifacts are found spatially clustered (Goren-Inbar et al. 2004). This study examines the presence of burned flint items, their spatial configuration, and the circumstances that contributed them to the archaeological horizons at Gesher Benot Ya'aqov.

1.3 The Acheulian Site of Gesher Benot Ya'aqov

Excavations at the Acheulian site of Gesher Benot Ya'aqov (GBY) were carried out along the east bank of the Jordan River in Israel (Figs. 1.2 and 1.3). From the early stages of excavation at the site, the occurrence of burned flint items within the archaeological horizons triggered an inquiry with regard to the circumstances in which these items had been burned. Burned flint items occur within various layers throughout the archaeological sequence; this sequence constitutes the major part of the 34 m long composite section that was compiled from the archaeological exposures, and mainly from the geological trenches (Fig. 1.3) (Goren-Inbar et al. 2000).

1.3.1 The Depositional Sequence

Subsequent tectonic activity on the transform fault of the Dead Sea Rift has resulted in the tilted form of the GBY strata (Figs. 1.4 and 1.5). The tilted archaeological layers are embedded within a generally fine-grained sedimentary sequence, with both organic-rich and carbonate-rich muds of autochthonous origin (Feibel 2001).

The thorough sedimentological analyses carried out by Feibel (2001, 2004, in prep.) have recognized three types of sedimentary settings associated with the archaeological layers included in this study. The first, consisting of gravelly or molluscan sands, derives from a storm event that deposited coarse sediments from the beach and near-shore environment on the shore face. Hominin activity (e.g., within the eight sequential archaeological levels of II-6 in Area B) was carried out following the formation of a stable surface on the beach, evidenced by the high degree of abrasion and polish on fragmented mollusks and by the sorting of the detrital clastics. Subsequent storm events covered each of the archaeological levels and "the freshness of most of the artifact edges implies that they accumulated well above the strand line, on the upper beach face" (Feibel 2001:137).

The second type of depositional setting (e.g., in Area C) is documented where the archaeological material is embedded at the interface between fine-grained offshore muds and a mollusk coquina. Following a drop in lake level, hominin activity was carried out on a land surface and the occupational surface was buried shortly afterwards due to a rise in

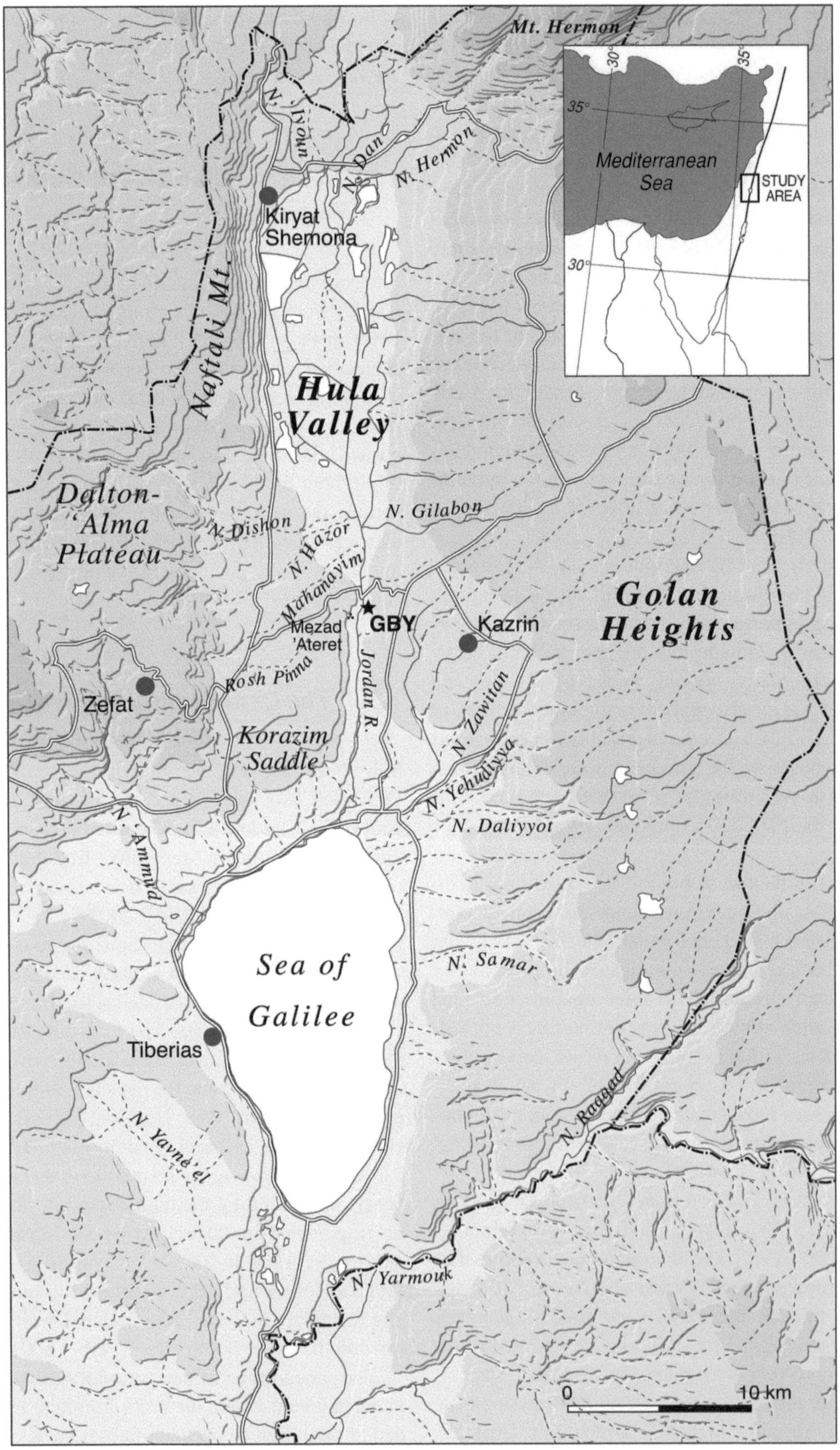

Fig. 1.2 Location map of the upper Jordan Valley (the Hula Valley and vicinity); the location of the GBY archaeological site is marked on the map (From Goren-Inbar et al. 2002b)

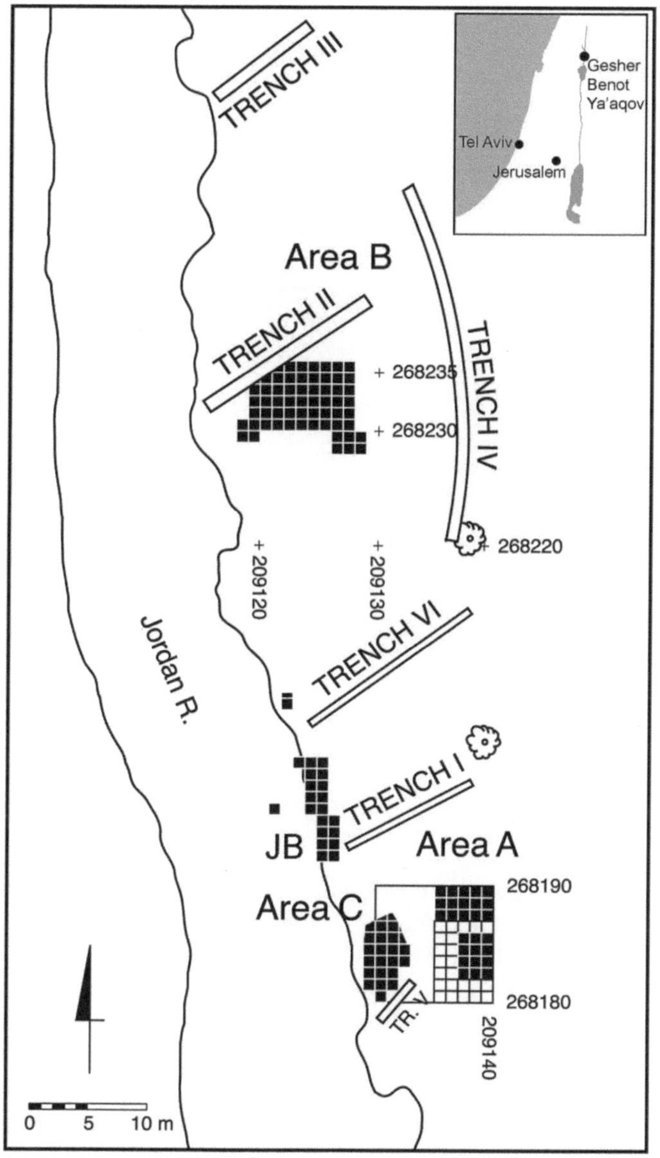

Fig. 1.3 Location map of archaeological excavation areas and geological trenches of GBY; coordinate numbers refer to the Israel Grid; point in inset shows the location of the study area within the Dead Sea Rift (33°00′28 N, 35°37′40 E)

lake level. This embedded offshore mollusks on top of the surface, incorporating some of the archaeological material into the coquina (Feibel 2001:137). This is considered to reflect "a short-term and discrete event burying the archaeological level in a transgression as is reflected in the unbroken and unabraded character of the bulk of the molluscan material" (Feibel 2001:139).

In the third association, the archaeological material is scattered within offshore muds, recording short-term regressions of the lake beach. The sediments occasionally demonstrate incipient soil features, and the archaeological material often provides the only indication of a brief episode of exposure; thus "only rarely was exposure prolonged enough to overprint the lacustrine muds with soil characteristics" (Feibel 2001:140).

The detailed sedimentological analyses of GBY provided a comprehensive reconstruction of the depositional environments for the different archaeological occurrences; by and large these analyses have documented rapid shifts in abundance of carbonate and organics, which are typical of a fluctuating lake margin environment (Fig. 1.6).

1.3.2 Chronology

The association between lithic artifacts and a variety of paleontological findings was observed following the first archaeological discoveries in the vicinity of GBY. The co-occurrence of extinct forms of freshwater mollusks (e.g., *Melanopsis aaronsohni* and *Viviparus syriacus*) in association with archaeological material provided a chronological marker of Early/Middle Pleistocene age, slightly older than the age suggested from the mammalian fauna (Picard 1963). This age was in accordance with the presence of characteristic Acheulian bifacial tools, and with the fact that they exhibit a typological and technological resemblance to African Acheulian assemblages (Stekelis 1960).

During the initial stages of the renewed excavations at the site (1989–1997), various characteristics of the lithic assemblages and fossil faunas (e.g., the recovery of a *Paleoloxodon antiquus* skull) placed the site in the general framework of the Middle Pleistocene, and an estimated date of ca 0.5 Ma was suggested based on the radiometric dating of a distant basalt flow, which is stratigraphically located beneath the site (Goren-Inbar et al. 1992).

Systematic sampling for a magnetostratigraphic study was carried out during the final season of excavations. Some 155 oriented samples have enabled reconstruction of the magnetic polarity history of 26 m of the composite section. The upper 17 m of the sampled sequence shows a northward declination and positive inclination, whereas the lower 9 m shows a southward declination and negative inclination (Goren-Inbar et al. 2000). These results illustrate that the GBY site consists of a reversed-polarity zone overlain by a normal-polarity zone. Given the biostratigraphical data as well as the characteristic lithic assemblages of the site, the only reasonable correlation of the polarity transition is to the Matuyama-Brunhes chron boundary, dated to 0.78 Ma (Goren-Inbar et al. 2000). The major archaeological levels included in this study occur above this boundary (i.e., 4 m for the base of II-6; ca. 13 m for the base of V-5). An attempt at combined ESR/U-series dating conducted on tooth samples from layers II-6 and II-7 yielded an estimated age of 652 ± 29 ka (Rink and Schwarcz 2005). This age confirms the assignment of the observed reversal in magnetic polarity to the Matuyama-Brunhes chron boundary.

Fig. 1.4 The northern face of Trench II; the major stratigraphic units of Area B are marked on the section; note the tilted position of the layers

The magnetostratigraphic reconstruction provides a single timeline within the cyclical sequence of GBY. The sedimentological analyses specified above have demonstrated a pattern of five second-order cycles of changing lacustrine facies recorded within a single first-order cycle that can likely be attributed to the ca. 20,000-year and 100,000-year Milankovitch cycles of global climate. Thus, the fact that the paleomagnetic boundary occurs between the first and second short-order cycles suggests that the age at the base of the composite section is ca. 0.8 Ma and the top of the section would date to 0.7 Ma (Goren-Inbar et al. 2000, 2002b; Feibel 2001). The duration of the entire depositional sequence at GBY is thus estimated as ca. 100,000 years, and it is assigned to OIS 18–20.

1.3.3 Cultural and Behavioral Issues

The GBY lithic assemblages were assigned to the Acheulian at very early stages of the excavations in the study area. The Acheulian technocomplex is identified by its characteristic large cutting tools (i.e., handaxes and cleavers), which first appear in East Africa at ca. 1.6 Ma (e.g., Kleindienst 1962; Roe 2001; Sharon 2007 and references therein). The Acheulian culture persisted until 0.3–0.25 Ma over a wide geographical range that includes Africa, the Iberian Peninsula, Europe, East-Central Asia, and the Levant (see Sharon 2007 for a detailed account on the geographical and chronological distribution).

The earliest human migrations into the Levantine Corridor occurred within the cultural complex of the Acheulian and involved the Levantine Corridor as a migration route out of Africa and into Eurasia. In the Levant, the earliest Acheulian occurrences are dated to 1.4 Ma at the site of 'Ubeidiya (Bar-Yosef and Goren-Inbar 1993; Shea 1999). As at GBY, the 'Ubeidiya assemblages exhibit varied typological and technological similarities to their African counterparts.

The Acheulian site of GBY thus displays the introduction of African traditions into the Levantine Corridor, which may reflect one of several repeated waves of human migration out of Africa (e.g., Saragusti and Goren-Inbar 2001). This is of great importance for this study, as the use of fire is often regarded as a triggering factor for the migration out of Africa, particularly with regard to the benefits bestowed by fire, which enabled the colonization of glacial Europe (e.g., Villa 2001).

The geographical and chronological position of GBY thus makes it a favorable site for examining the issue of early use of fire. In addition, the systematic procedures of excavation have enabled the retrieval of unique botanical remains, as well as faunal and lithic assemblages, that can be used in the attempt to reconstruct varied environmental, cultural, and behavioral aspects of the Acheulian occupation at the site. While the archaeological materials retrieved from the site are currently

Fig. 1.5 General view of the excavations at Area B (1996 season); view to the south from Trench II

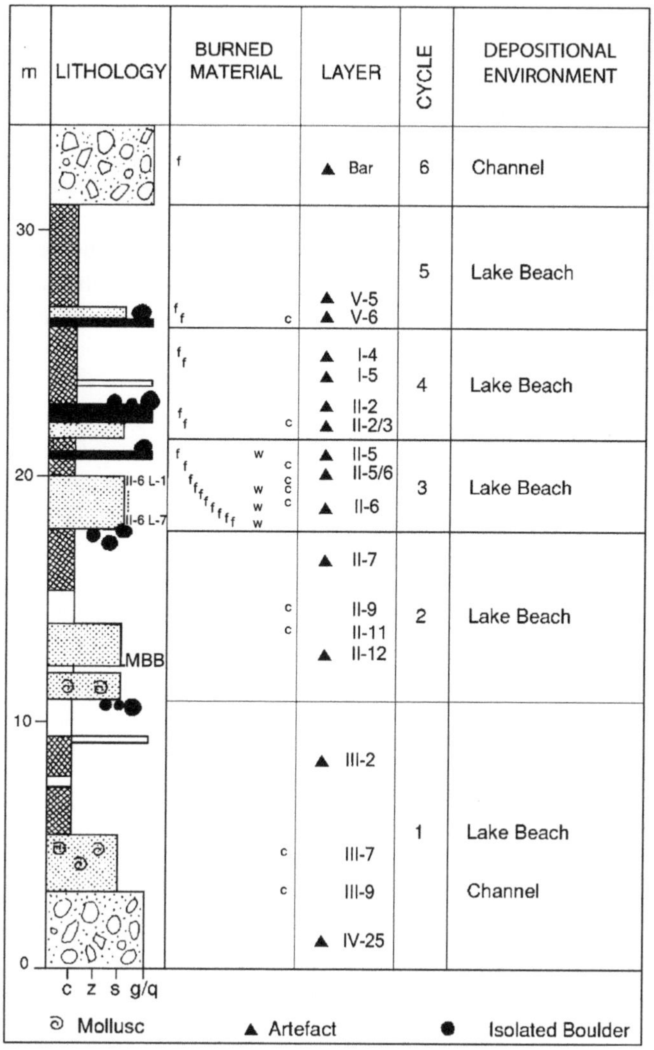

Fig. 1.6 Composite section of the GBY sequence with correlation to major stratigraphical units and cyclicity of the sequence (c = clay, z = silt, s = sand, g/q = gravel/coquina) with presence of burned material (w = burned wood, c = charcoal, f = burned flint); location of the Matuyama-Brunhes chron boundary (MBB) is marked on the section (Modified after Feibel 2001)

under study, preliminary examination of diverse proxies has contributed to our knowledge of the hominins[3] of GBY.

1.3.3.1 Lithic Production

The lithic assemblages of GBY exhibit a distinct pattern of raw material selection in which specific morphotypes are associated with specific raw materials. Limestone and flint pebbles, most likely collected from wadi beds and terraces in the immediate area of the site, were used for the production of cores, flakes, and a variety of flake tools (flint) as well as chopping tools and percussors (limestone). Bifacial tools are modified primarily on basalt, which is the dominant raw material in the vicinity of the site and is available in the form of flows, boulders, cobbles, and pebbles (Goren-Inbar and Saragusti 1996; Goren-Inbar et al. 2000).

The preference for basalt for the production of bifacial tools is fairly constant throughout the stratigraphic sequence. However, other components of the lithic assemblages (i.e., cores and core tools, flakes and flake tools) demonstrate variability in the frequencies of different raw materials (Goren-Inbar et al. 2000). Similarly, the representation of different lithic components (e.g., tools, cores, and waste products) is variable throughout the archaeological sequence, probably reflecting "different hominin activities along the shores of the paleo-Hula Lake over time" (Goren-Inbar et al. 2000:947).

Various patterns within the knapping reduction sequence, which may account for the observed variability of certain lithic components, were identified through extensive knapping experiments

[3]Hominin skeletal remains were not discovered in recent excavations at GBY. Two human femoral diaphyses were found in the assemblage of faunal remains originating from the early excavations at the site (see Geraads and Tchernov 1983; Goren-Inbar and Belitzky 1989).

(Madsen and Goren-Inbar 2004); these have underlined the role of selection and transportation of good-quality basalt by the GBY hominins. The experiments have further demonstrated that the flaking of heavy (i.e., giant) basalt cores may have commenced at the outcrop, as the various size categories of the artifact classes, as well as the under-representation of certain artifacts in the archaeological assemblages, clearly point to the import and export of products into the site (Madsen and Goren-Inbar 2004; Goren-Inbar and Sharon 2006). Thus suitable large flakes were sorted and selected as blanks for future preparation of handaxes, cleavers, and other flake tools (e.g., massive scrapers: Goren-Inbar et al. 2008).

Knapping at GBY, including the modification of large blanks into bifacial tools, involved the use of the soft-mode technique, i.e., knapping with soft percussors made from fallow or giant deer antlers or massive bone parts. This is suggested by both the experimental studies and various morphological characteristics of the flake component (Sharon and Goren-Inbar 1999; Madsen and Goren-Inbar 2004). Overall, the design and quality of the bifacial tools, which are rooted in long in-depth planning, indicate that the operative competence shown by the GBY hominins in tool making is comparable to that of modern humans (Goren-Inbar 2004; Madsen and Goren-Inbar 2004).

1.3.3.2 Animal Exploitation and Processing

The site of GBY exhibits sequential occurrences of hominin activities that occurred in proximity to the various resources available in the lakeside environment. Apart from the availability of large amounts of raw material (discussed above), the presence of fresh water and its attraction for large game (e.g., elephant, large bovids, hippopotamus, rhinoceros, etc.) and small game (e.g., gazelle, wild boar) were favourable conditions for the recurrent occupations observed at GBY.

The faunal assemblages of GBY reveal a mixture of Asian and African components and demonstrate the exploitation of diverse taxa, the remains of which occur within the archaeological horizons and often exhibit features characteristic of human modification (e.g., cut marks: Rabinovich et al. 2008). Examples of such modification can be seen in the battered skull of *Paleoloxodon antiquus*, which was found with numerous associated fragments along with lithic artifacts of various sizes. The skull, unearthed within a distinct archaeological level (II-6 L-1), is considered to reflect the terminal dismemberment stage of an elephant hunt (Goren-Inbar et al. 1994).

The detailed taphonomic studies of the bone assemblage have revealed that the hominins of GBY had a specific knowledge of animal anatomy. The butchering procedures, evident in the repetitive locations of cut marks and conchoidal fractures, are indicative of a wide anatomical knowledge (Rabinovich et al. 2008).

Signs of burning have not been identified on bones from GBY. It seems that the fossil state of the bones, which has been affected by the anaerobic waterlogged environment of the site (i.e., the mineral replacements and the dark patina that characterize the faunal assemblage), does not permit the application of conventional techniques to identify burning on the bones.

1.3.3.3 Plant Exploitation

The waterlogged environment of GBY has enabled the preservation of a unique paleobotanical assemblage, which includes wood, bark, seeds, fruits, and pollen. The various identified taxa (Melamed 1997, 2003; Goren Inbar et al. 2002b) represent a Mediterranean submerged bank and dry-land vegetation and are indicative of the Pleistocene paleoenvironment of the Hula Valley (Goren-Inbar et al. 2000).

The role of hominins in the formation of the paleobotanical assemblages is uncertain; some of these remains could have been transported into the site by the hominins, and others might have been naturally embedded within the sediments. Yet the presence of various edible species within the distinct archaeological horizons (e.g., wild grape [*Vitis sylvestris*], wild olive [*Olea europaea*], and jujube [*Ziziphus*]), in association with lithic artifacts and modified animal bones, is suggestive of the use of extensive and rich foraging resources.

Human use of floral resources can be found in the unique wooden artifact found at GBY. The fragmented artifact was made from willow (*Salix*) and exhibited two surfaces – a single polished flat surface and an opposing convex surface (Belitzky et al. 1991). Another example is the association of the assortment of edible nut species (i.e., water chestnut [*Trapa natans*], prickly water lily [*Euryale ferox*], wild almond [*Amygdalus communis*], pistachio [*Pistacia* sp.], and oak [*Quercus* sp.]) with the various pitted stone anvils from GBY. This is interpreted as evidence for nut cracking, which enabled the GBY hominins to exploit the high nutritional value of the nuts (Goren-Inbar et al. 2002a). In the framework of the current study, it is interesting to note that some of the nut species of GBY (e.g., wild almond and oak) contain toxic substances that can be reduced by roasting. This reflects one of the advantages provided by the use of fire for the expansion of dietary resources (see Chapter 4 for further discussion).

The paleobotanical assemblages of GBY have contributed to the study of fire use at the site. Among the wood fragments larger than 2 cm, 1.41% (n = 13) of the specimens from all the excavated areas were identified by microscopic analysis as burned wood. Ranging in length from 2.3 to 10.1 cm, these specimens originate in six archaeological horizons; two taxa were identified as wild olive (*Olea europaea*) and Syrian ash (*Fraxinus syriaca*) (Goren-Inbar et al. 2002b: Table 20).

Signs of burning were also observed on fruits and small wood fragments; of the fruits and small wood fragments, 4.47% (N = 426) were identified as burned and comprise various identifiable taxa: Syrian ash (*Fraxinus syriaca*), wild olive (*Olea europaea*), Greek silk-vine (*Periploca graeca*), willow (*Salix* sp.), oat (*Avena* sp.), wild grape (*Vitis sylvestris*), mesquite (*Prosopis*), bedstraw (*Galium* sp.), wild barley (*Hordeum spontaneum*), and goatgrass (*Aegilops geniculata*) (Goren-Inbar et al. 2004).

The various fields of study analyzed in the framework of the GBY excavation project demonstrate that the hominins had a comprehensive knowledge of their environment. The exploitation of the various available natural resources enabled them to use the site as a base for a variety of activities, including lithic production, animal butchering, and food consumption. Controlled use of fire may well be another example of the behavioral complexity of these hominins.

Chapter 2
Framework of Research

The discussion in the previous chapter of the evidence for the early use of fire has demonstrated the complexities involved in identifying the early stages of human control over fire and emphasized the need for a comprehensive approach to their study. The presence of burned flint items at the site of GBY provides a unique opportunity to investigate this pressing issue with a different methodological approach. This methodology draws on a variety of ethnographic, archaeological, and ethnoarchaeological studies, which generally suggest that small lithic products can be used as spatial indicators for a variety of activities, including the use of fire in the form of hearths. The theoretical foundations of this approach are presented in this chapter, which integrates the different components of the research program.

First, this chapter presents the *research objectives* of this study. Although these were initially designated for the analysis of burned flint items from GBY, they contribute directly to a much wider theme, the examination of hominin behavior and technological skills, which is also embedded within the objectives of this study.

Following the research objectives, the principles of the methodological approach are presented. The various research assumptions, as well as the construction of the applied methodologies, all draw on an extensive theoretical framework of *hearth-related spatial patterning*, which is discussed in the second part of this chapter. An understanding of these theoretical foundations is essential before approaching the *research hypotheses* and *methodology* of the study, presented at the end of this chapter.

2.1 Research Objectives

The initial objective is to report on the presence of burned flint items in the archaeological horizons of the Acheulian site of GBY. The burned flint items, which occur throughout the stratigraphic sequence in varying frequencies and in diverse spatial settings, are presented and analyzed in this work.

The burning of these flint items may have been the outcome of natural[1] or anthropogenic fire. Accordingly, the second objective of this study is to examine the possibility that these items are the result of anthropogenic fire (i.e., fire used by hominins). In doing so, we may establish evidence for the use of fire at GBY as early as 790,000 years ago. In order to achieve this goal, it is essential to obtain a reliable means of distinguishing an anthropogenic fire from a natural one. This is accomplished throughout the analysis of the spatial configuration of burned flint items, thoroughly discussed in the following parts of this chapter.

Considering the various advantages provided by fire, establishing evidence for the use of fire at GBY is of great importance for the evaluation of hominin behavior at the site. Thus, the third objective of this study is to inquire into different behavioral and technological aspects of the use of fire. These include the role of fire in the different occupation episodes of the site (i.e., the presence or absence of fire use in relation to different activities), and the apparent frequency of the use of fire throughout the various occupation episodes recorded at the site (i.e., is the use of fire a unique phenomenon or a routine practice).

2.2 Hearth-Related Spatial Patterning

The following discussion presents the theoretical foundations of this study, which generally suggest that the spatial patterning of a variety of activities, including the use of fire in the form of hearths, can be implied by lithic waste products of small size.

Human activities are spatially patterned and the fact that humans tend to carry out a vast range of activities in close vicinity to hearths is widely documented. The hearth assembles the social group and around it is the area in which social interactions, tool production, food processing,

[1] A thorough discussion of fire ecology, and specifically of the probability of natural fire at GBY, is included in Chapter 4.

N. Alperson-Afil and N. Goren-Inbar, *The Acheulian Site of Gesher Benot Ya'aqov Volume II: Ancient Flames and Controlled Use of Fire*, Vertebrate Paleobiology and Paleoanthropology, DOI 10.1007/978-90-481-3765-7_2, © Springer Science+Business Media B.V. 2010

food consumption, and ritual ceremonies are carried out (e.g., Yellen 1977; Binford 1983, 1998; Spurling and Hayden 1984; Galanidou 2000, 1997). While numerous activities (e.g., social interactions) leave no tangible evidence for us to uncover, other activities (e.g., tool making and food processing) contribute directly to the formation of the archaeological record. Brooks and Yellen (1987) defined *procurement*, *processing*, *consumption*, and *manufacturing* as principal "*debris-generating*" behaviors. The latter involves the manufacturing of artifacts and is strongly associated with hearths (Brooks and Yellen 1987:82).

Hearths not only serve as spatial spots of accumulation but also influence the patterns of distribution of certain size groups of the assemblage. Binford (1978, 1983) suggested that the formation of certain spatial patterns during work around a hearth appears to be universal. More specifically, the distribution of debris often displays two concentric zones around the hearth: the *drop zone* in proximity to the hearth, where small fragments of bone/stone are left in situ (*residual primary refuse* in the terminology of Schiffer 1972, 1987), and the *toss zone*, an area further away from the hearth to which the larger debris is tossed (*secondary refuse* in the terminology of Schiffer 1972). Thus the area closest to the hearth is likely to display high quantities of small in situ refuse.

2.2.1 Small In Situ Refuse

The fact that small items are left in their original location while large items tend to be removed was reported as early as 1961 in Green's pioneering study of discard patterns (Green 1961:91). Notwithstanding, spatial analysis studies often concentrate on the larger refuse and features, despite the fact that "…the data most likely to be informative … are very small refuse items, such as chipping debris, small bone fragments, and plant macrofossils, which will often be found in primary context" (O'Connell 1987:104).

Smaller refuse is more likely to be found in situ for several reasons. Small items are less visible and are more likely to be missed during refuse clearance and preventive maintenance of the activity area (e.g., DeBoer 1983; Schiffer 1987), their small dimensions make them less hazardous (e.g., Hayden and Cannon 1983; Clark 1991), and they are more prone to trampling and thus penetrate deeper into the occupation surfaces (see DeBoer 1983 for a detailed discussion).

The fact that small refuse is more likely to be left in situ than large refuse is known as "McKellar's principle" (first published in Schiffer 1976:188). McKellar's work on the litter of the University of Arizona campus indicated that there is a critical size factor in refuse disposal patterns. She had found that items above 9 cm were consistently tossed into trash cans, while smaller items were left behind as primary refuse (Rathje 1979:10; Schiffer 1976:188, 1987:62). McKellar's principle has been confirmed in a variety of ethnoarchaeological studies (e.g., Schiffer 1987:62 and references therein, Stevenson 1991 and references therein). However, while the general principle has been widely adopted, no conventional limit has been defined as the critical size factor. In other words, what is considered small?

One extreme would be particles smaller than 1 mm (*microdebitage* in the terminology of Fladmark [1982], referring only to stone knapping products). Under a microscope, microdebitage can be further divided into *microflakes* and *microchunks* (Vance 1987). A maximum size of 2 mm, *microartifacts* in the terminology of Stein (Dunnell and Stein 1989; Stein and Teltser 1989, referring to all archaeological residues), has also been suggested. These microartifacts have been found to be significant in the study of both natural (see Dunnell and Stein 1989) and cultural (e.g., lithic manufacturing and discard: Hull 1987; duration of occupation: Simms 1988) formation processes. Other studies set the limit at 2.5 mm (Metcalfe and Heath 1990), 6 mm (Austin et al. 1999), 10 mm (Nadel 2001), 20 mm (Alperson-Afil and Hovers 2005), 25 mm (DeBoer 1983) or 50 mm (O'Connell 1987).

Despite the variability in scale, the various studies all share the view that small items are essential components in the reconstruction of site structure and are optimal indicators of activity areas (e.g., Hayden and Cannon 1983:134; Schiffer 1987:94; Simms 1988:208; Cessford 2003:3).

In conclusion, ethnographic observations have laid the foundations for site structure reconstruction, which is based on the recognition that the association between features (e.g., hearths) and the spatial distribution of artifacts can provide the contextual framework of artifact concentrations (Simek 1984). Consequently, in attempting to reconstruct the formation process of hearth-related spatial patterns, we can draw on the following inferences:

1. A wide range of activities is carried out in close proximity to hearths.
2. Hearths are spatial spots of refuse accumulation.
3. Small refuse is more likely than large refuse to be left in situ.
4. Hearths are thus likely to display dense concentrations of small refuse.

Archaeological evidence of similar hearth-related discard patterns has been reported as early as the Middle Paleolithic (e.g., Vaquero and Pastó 2001) and from a variety of archaeological settings. These include open-air sites (e.g., Gilead 1980; Hietala 1983; Gilead and Grigson 1984; Goring-Morris 1988, in prep.; Leesch et al. 2005; Sergant et al. 2006), rockshelters, and cave sites (e.g., Galanidou 1997; Vaquero and Pastó 2001), in all of which the hearths are readily identifiable features.

2.2.2 Phantom Hearths

We are often required to characterize artifacts or features recovered from archaeological contexts. Hearths, however, are features of all contemporary hunter-gatherer societies, and when found in such contexts they exhibit high variability of construction methods, size, and functions. The lack of a clear archaeological definition of a hearth appears to result from their universal contemporary occurrence, as well as from their apparent variability. The recent ethno-geoarchaeological project carried out by Mallol et al. (2007) characterized different sedimentary aspects of Hadza hearths through soil micromorphology. Their study provides examples of a variety of Hadza hearths and illustrates the variability in construction, morphology, intensity, and function of the hearths. Galanidou (1997) provides a summary of ethnographic examples of hearths used by hunter-gatherers and horticulturists in caves and rock shelters; again, a high degree of variability is recorded for the types, number, and functions of the hearths. These case studies also emphasize the notion that a hearth is not necessarily a built (e.g., stone-lined) feature; out of nine case studies, five groups use open hearths, three use stone-lined or log-lined hearths, and yet another uses open hearths that are occasionally lined with stones (Galanidou 1997:141–144). Ethnographic data thus suggest that open hearths involving no construction are more common than hearths requiring the excavation of shallow or deep pits, lining with stone or wood, or structuring of any sort: "Hearths made directly on the underlying substrate without any particular previous preparation appear to be a well established transcultural phenomenon … the demarcation of combustion zones with stones is limited in the world of modern foragers" (Meignen et al. 2007:103).

Thus, if we were to define a hearth we would not include the building or structuring of the combustion area as a basic element. It seems that the only common feature of all hearths is the simple fact that people intentionally burn fuel in order to produce a fire. Accordingly, an archaeological definition of a hearth will specify that a hearth is a combustion area, variable in structure, size, and depth, which preserves the remains of burned materials. In his "Dictionnaire de la Préhistoire" Leroi-Gourhan suggested the following definition of a hearth: "Dans la terminologie ancienne, est souvent synonyme de *couche archéologique*, celle-ci se révélant par un sédiment sombre comportant des charbons de bois et des foyers au sens strict" (Leroi-Gourhan 1988:405). According to this definition, the hearth will exhibit discoloration (dark sediments) and charcoal will be preserved. Schiegl et al. (1996) suggest another definition that similarly depends on the state of preservation: "Good field evidence for the use of fire is the presence of well preserved hearths. Such hearths are usually round or oval-shaped and often have an upper layer composed of light coloured minerals, a lower layer rich in charcoal, and a substrate of reddened sediment" (Schiegl et al. 1996:763–764).

These descriptions, however, appear to suit the definition of a "well-preserved hearth" better than that of a "hearth". The ethno-geoarchaeological study of Mallol et al. (2007) mentioned above demonstrated that the preservation of combustion features is not a straightforward issue, particularly in open-air sites: "Micromorphological results suggest that the anthropogenic signature of open air combustion structures can be detected depending on the rates of sedimentation and the impact of postdepositional disturbance factors … If the rates of sedimentation are low, leading to erosion, the remains of an ephemeral open air fire are likely to disappear" (Mallol et al. 2007:2050). Similarly, discoloration of sediments around and beneath the hearth depends on a variety of factors (e.g., fuel used, soil moisture, chemical variations in sediments) and requires favorable depositional and post-depositional conditions in order to be preserved in the archaeological record (Bellomo and Harris 1990; Canti and Linford 2000; Linford and Canti 2001; see also the discussion in Section 1.2.1).

In summary, it is evident that the definition of a hearth varies in terms of sedimentological setting, intensity, size, fuel used, structure, and function. These variables will eventually dictate the archaeological appearance of these features, i.e., whether hearths will exhibit a stone lining, whether ash and/or charcoal will be preserved, or whether discoloration of the sediments will occur. Consequently, as in the ethnographic record, the archaeological occurrences of hearths are extremely variable and uneven, and hearths are independently defined for each site.

Examples from the Middle Paleolithic include Abric Romani, where hearths were identified "by the presence of homogenous lenses of ash and charcoal, and thermal alterations of the underlying surface" (Vaquero and Pastó 2001:1212). At Kebara Cave, hearths appear "… in different forms, most often as lenses consisting of black and white layers of varying dimensions but also as ashy white accumulations; grey sediments composed of consolidated aggregates of ashes and black charcoal; zones of consolidated grey ash; and alternating thin, grey-white and black layers over large areas" (Meignen et al. 2007:93). At Grotte XVI (Dordogne, France), analyses of the thick ash deposit clearly defined the presence of hearths: "However, none of the more complex preparations – rock lining or excavated pits – associated with later Upper Paleolithic features have been discerned …" (Rigaud et al. 1995:911). At Site C in Belvédère, where both charcoal and burned flint artifacts were recorded (Stapert 1990), the location of the hearth was nonetheless recognized in the center of concentrations of burned flint, as the charcoal was probably "… carried away by flowing water after abandonment of the site. It seems that the flowing water did not have a strong erosive effect, because it left the flint concentration, including many tiny chips, in place" (Stapert 1990:5).

As the archaeological appearance of hearths is variable in color, size, contour, depth, and the use of stones for construction, it is difficult to generate an archaeological definition that suits these features. However, since hearths serve as focal points for activities, they display areas of refuse accumulation, specifically small refuse. These patterns can easily be identified when we examine sites in which the hearths are well preserved.

In this study, however, we are concerned with *phantom hearths* that display no directly observable features. Leroi-Gourhan's definition of *structures latentes* established the approach to such archaeological features, namely that these can be discernible through observable patterns of artifacts' spatial distributions (Leroi-Gourhan and Brézillon 1972).

Considering the hearth-related spatial patterning discussed above, we assume that clusters of debris, specifically small burned debris, are indicators of the locations of hearths and are defined as *phantom hearths* – features that lack other observable traits (e.g., structuring, discoloration of sediments, ash, charcoal). If we were to pursue the locations of the hearths, we should be able to trace them in the center of these concentrations. At the Middle Paleolithic occupation at Belvédère, clusters of burned artifacts suggested the presence of hearths in the centers of these concentrations (Stapert 1990). At the Magdalenian sites of Champréveyres and Monruz in Switzerland, hearths are characterized by various amounts of cobbles, stone slabs, and extremely abundant and well-preserved wood charcoal (Leesch et al. 2005). Regardless of the remarkable preservation of these sites, the spatial distribution of burned flint microartifacts has proved to be an optimal indicator for the precise location of the hearths, illustrating "...the legitimacy of mapping the burned flint chips to locate the combustion areas" (Leesch et al. 2005:7). Similarly, at the Mesolithic site of Verrebroek in Belgium, the patterns illustrated in the archaeological data were supported by various experimental studies that suggested that "simple surface hearths can be localized quite accurately on the basis of the distribution of severely burnt or overheated chips (2 mm–1 cm)" (Sergant et al. 2006:1006). In addition, it has been suggested that small items exhibit higher frequencies of burning than large items. Similarly to the observed effects of fire on bones (e.g., Stiner et al. 1995; Villa et al. 2002, 2004), this pattern may result from the fact that fire fractures and cracks material into smaller pieces, resulting in higher frequencies of burning amongst small items. A recent experimental study (Sergant et al. 2006) has demonstrated strong fragmentation of flint artifacts caused by burning. In this experiment "the initial 143 artifacts (larger than 1 cm) and 530 chips were shattered to 240 artifacts (larger than 1 cm) and 3419 chips, i.e., a multiplication with factor 1.7 and 6.45" (Sergant et al. 2006:1002). Since large quantities of small-sized burned flint items are expected to be found, large enough samples can be available for spatial analysis.

Thus, based on the above observations, spatial clustering of burned material, specifically small burned flint items, is considered in this study to be the main criterion in the identification of anthropogenic fire. The various possible spatial configurations of the small burned flint items at GBY are embedded within the different research models, specified below.

2.3 Research Hypotheses

The general hypothesis of this study is that the presence and spatial configuration of burned flint items can be used to identify anthropogenic fire in the attempt to establish evidence for hominins' early use of fire. Accordingly, the research hypotheses integrate both of these aspects – the identification of burning damage on flint and the characterization of their spatial clustering. However, as in any spatial analysis study, we are compelled to assume adequate preservation of the original spatial configuration of the archaeological occurrences. Thus, before addressing the hypotheses concerning the effects of fire on flint and the spatial patterning of anthropogenic fire, the following short discussion concerns the various sedimentological and taphonomic considerations suggestive of the preservation of the original spatial configuration of the archaeological remains at GBY.

2.3.1 Taphonomic Considerations

In the framework of this study, two different factors support the assumption that the observed distribution patterns represent the original configuration of the archaeological material. First, the archaeological occurrences of GBY are recorded in a lake shore environment, in which the oscillating water level of the lake allowed the rapid sealing of the archaeological material (Feibel 2001). Accordingly, various taphonomic observations suggest that post-depositional processes had a limited effect on the original location of the archaeological material (e.g., Goren-Inbar et al. 1994, 2002b, 2004; Ashkenazi et al. 2005; Goren-Inbar and Sharon 2006; Rabinovich et al. 2008, in press).

Secondly, the major components of this study are small flint items. As previously discussed, such small items tend to remain in their original location for a variety of reasons and are thus particularly reliable components in any spatial analysis. Despite this, in a lake shore environment such as that of GBY microartifacts may be subjected to particular taphonomic phenomena associated with lake margin processes.

Experimental studies of such processes carried out by Morton (1995) provide valuable data on the association between artifact weight and transport mode. It has been suggested that during lake transgression or regression events, heavier artifacts tend to subside with the sediment, while the "… lighter artifacts would either be transported downshore or downslope" (Morton 1995:77). In such a case, we would expect to find that smaller items do not cluster together but are rather arranged in a linear distribution along the presumed shore line. Lake margin environments can also spatially rearrange material in a non-linear manner and form clusters of denser concentrations. Natural features (e.g., fallen trees, bushes, or beach cusps) can provide obstructions and act as accumulators in a lacustrine environment. The comprehensive experiments carried out by Morton (1995) have demonstrated that in such cases, during transgression and regression of the lake, the heavier artifacts remain buried with minimal disturbance while the smaller ones became trapped and then buried at the obstruction. However, the "… complete absence at the obstruction of any artifacts over 10 grams …" (Morton 1995:120) led Morton to the following conclusion: "The recognition of a 'hydraulic jumble' (Isaac 1984) has been an important pursuit as an alternative hypothesis for the formation of archaeological sites. The experiments outlined have showed that these accumulations can indeed occur, but not without an obvious 'fingerprint'. Far from being a 'jumble', these sorts of archaeological concentrations would consist of well-sorted artifacts, all within a specific weight category" (Morton 1995:122).

Morton's experiments emphasize the strong relationship between artifact weight and wave energy. Accordingly, in low-energy environments, like that of GBY (Feibel 2001), the varied taphonomic phenomena that are often illustrated in lake margins will not necessarily occur: "Since a combination of random hydrodynamics and proximity to other particles causes artifacts to be braced into the sediment, they are quickly covered over or the edges become blanketed in sediment. This causes the artifacts to assume a lower, less resistant profile and hastens sedimentation. It is this sort of hydrodynamics that ensures that in a lake margin … it is possible that all smaller flakes and debitage are not removed" (Morton 1995:177).

In the framework of this study, we can benefit from these observations while analyzing the distribution patterns of flint microartifacts. We can thus assume that if the archaeological occurrences of GBY were significantly subjected to lake margin processes that rearranged the spatial configuration of the archaeological material, then flint microartifacts:

1. Will not exhibit clustering but rather a linear distribution along a presumed shoreline.
2. Will exhibit clustering; however, since size sorting is involved, they will not be associated with larger artifacts.

2.3.2 The Effects of Fire on Flint

A basic assumption in this study is that the identification of particular burning damage on flint items from an archaeological occupation provides sufficient evidence for the presence of fire. This assumption is based on a variety of experimental studies that have demonstrated the different effects of fire on flint.

Exposure to fire changes the mechanical properties of lithic material. Experimental studies, carried out mostly on flint or chert, demonstrated that exposure to high temperatures (~350–500°C) causes macroscopically identifiable (i.e., visually observable) alterations such as discoloration, potlid fractures, crazing, and fragmentation (Purdy and Brooks 1971; Purdy 1975, 1982; Julig et al. 1999; Sergant et al. 2006).

Early experiments in heating of silica minerals recognized that processes of expansion and contraction, which occur when materials are heated or cooled rapidly, result in explosions of the heated material (Crabtree and Butler 1964). However, the identification of specific features of burning damage and the particular means by which they are formed was first presented by Purdy (1975; Purdy and Brooks 1971). Her experiments have shown, for example, that "potlids always occurred during the heating process, never during the cooling process; thus they must be a result of expansion" (Purdy 1975:136).

Potlids are small, typically "bowl-shaped" pieces of lithic material that are exfoliated from the surface. Their exfoliation creates a depression (i.e., a potlid fracture) in the artifact from the size of a pinhead to larger (DeBano et al. 1998:271).

Recent experiments have demonstrated that only those artifacts that are in direct contact with the fire and heated to a temperature above 300°C will eventually show heat damage (Sergant et al. 2006). These experiments subdivide fire-damaged flint artifacts into three classes: (1) weakly burned: few traces of heat damage, except for a weak reddish shine and a few isolated cracks; (2) moderately burned: more visible heat damage, such as potlid fractures, cracks, and color changes; and (3) heavily burned (overheated): display total dehydration resulting in a white to gray discoloration (Sergant et al. 2006:1000).

In sum, only direct exposure of flint to fire results in visible heat damage, and heat damage is diversified and includes a variety of features. In the attempt to identify the presence of fire at a site as ancient as GBY, we chose to be extremely cautious and consider only items that are unquestionably burned. Therefore, the identification of burned flints had to rely on features that are clear and unique to exposure to fire. Of the various heat-damage patterns, potlidding is the most distinctive feature.[2] Thus, we assume that the occurrence of

[2]The use of discoloration of flint as a distinctive feature for the identification of burning is not a reliable measure for the flints of GBY; embedded within the waterlogged sediments, the majority of flint items are darkly patinated.

potlid fractures on flint items from the site of GBY is sufficient evidence for the presence of fire.

2.3.3 Patterns of Anthropogenic Fire

The identification of burned material at an archaeological occupation attests to the presence of fire; however, it is necessary to ascertain that this fire is anthropogenic rather than natural.[3] In this study, following the theoretical foundations discussed above, it is assumed that anthropogenic fire, in the form of hearths, will not damage flint items throughout the entire occupation surface but rather result in relatively small but non-random frequencies of burned items that are spatially clustered. Consequently, the spatial patterning of burned flints can confirm the presence of anthropogenic fire at an archaeological occupation.

The attempt to identify clusters of burned flint items is, however, accompanied by true complexities. When defining spatially discerned activities, we usually assume an association between particular tasks and specific archaeological material (e.g., an animal processing spot should exhibit spatial aggregation of worked bones, stone knapping areas should similarly display denser areas of lithic debris, etc.). However, when attempting to identify clusters of burned flints, we are actually examining items that spatially derive from a larger bulk of the flint component of the analyzed surface. This is of great importance, since the flint component may *a priori* be spatially clustered; thus, in a case where flint knapping was confined to a specific location, the original spatial patterning of the flints will exhibit clustering. A random, or uniform, burning pattern upon this will also appear clustered, so that any fire, whether anthropogenic or natural, can result in clustering of burned flints. Thus, where the unburned and the burned flints entirely overlap (spatially) each other, we cannot rule out the possibility of a natural fire. Based on these notions, several models are examined regarding the presence and distribution of burned flint items in each of the archaeological horizons analyzed in this study. These models particularly emphasize the spatial configuration of the burned flint microartifacts with respect to the unburned ones:

Model 1: Burned flint items are not present in the examined horizon; thus there is no evidence of fire, whether natural or anthropogenic; *Model 2*: Burned flint items are present in the examined horizon. However, the burned and unburned flint items are distributed identically; thus there is evidence for the presence of fire at the site but we cannot rule out the possibility of a natural fire, deforming flint wherever it occurs; *Model 3*: Burned flint items are present in the examined horizon. In addition, these burned flints occur in distinct clusters, whereas the burned and unburned items are not distributed identically. Thus there is evidence for the presence of fire at the site, and this fire has deformed flint in specific localities that do not entirely coincide with the original distribution of the unburned flint. This resulted in spatially discerned clusters of burned material that can be confidently interpreted as the remnants of anthropogenic fires.

These models will be examined for each of the analyzed archaeological horizons.

The means by which these models are evaluated are presented in the following section, which incorporates the various methodological procedures applied in this study.

2.4 Methodology

The previous sections of this chapter have laid the foundations of the methodological approach applied in this study: to examine the presence and spatial distribution of small burned flint items in order to identify possible clusters of burned material. The following section presents a detailed methodological account of the various procedures involved in the identification, analyses, spatial plotting, and spatial analyses of the burned flint items from GBY and their associated lithic assemblages.

The different methods comprise two stages, the second dependent on the first. The first involves the compilation, analysis, and preparation of the data, which eventually enabled its transformation into a geographical information database. The second includes the various tools used for the spatial display of the data and analysis of the observed spatial patterns.

2.4.1 Excavation Methods and Provenance Recording

Excavations at the site of GBY were carried out in three main areas, all located on the eastern bank of the Jordan River (Fig. 1.3). A horizontal 1 × 1 m grid was constructed above the excavated surfaces, corresponding to the coordinates of the Israel grid.

Excavation was conducted along the strike and dip of the layers with the aim of exposing the tilted archaeological horizons laterally (Goren-Inbar et al. 2002b: Figures 4 and 5); this procedure enabled the detailed representation of the spatial organization of each occupation surface. The standard unit of excavation was thus the tilted projection of a horizontal 1 m^2 grid square. Each horizontal grid square was further

[3] As previously noted, a thorough discussion of fire ecology, and specifically of the probability of natural fire at GBY, is included in Chapter 4.

subdivided into four 0.5 × 0.5 m sub-squares and excavated in spits that covered the area of one sub-square to an average depth of 5 cm.

Once exposed, the surface (i.e., the living floor) was drawn and items were retrieved with a full spatial reference (X, Y, and Z); these "coordinated pieces" consist mostly of items larger than 2 cm (i.e., macroartifacts). Other items retrieved during excavation, the "uncoordinated pieces", were labeled according to the spatial reference of the spit (i.e., excavated unit/sub-square, and an elevation range). Such items can thus be located with an exactitude of 0.5 × 0.5 × 0.05 m.

In addition to material retrieved during excavation, the entire excavated volume of sediments embodying the archaeological horizons was wet-sieved during field work[4] using a 2 mm sieve. The wet-sieved sediments were then bagged with their recorded spit location and transported to the Institute of Archaeology for further analysis. Sorting of the sieved sediments yielded rich and varied assemblages, such as fruits, seeds, grains, bones and teeth of micromammals, fish, and crabs, and specks of charcoal. Most of the small lithic items, which are the main evidence on which this study is based, were retrieved through this procedure. These include all stone items (basalt, flint, and limestone) that range in size from 2 to 20 mm (henceforth microartifacts). As the wet-sieved sediments were retrieved from the field with their recorded spit location, these microartifacts can be located with an exactitude of 0.5 × 0.5 × 0.05 m.

2.4.2 Analyzed Samples

As this study examines the spatial distribution of burned flint items, the spatial recording of the archaeological material is fundamental. Thus, items for which the spatial data are incomplete (i.e., retrieved items for which the spatial record is lacking) are not included in the analyses. Accordingly, the various illustrations as well as the corresponding summary tables of the lithic inventory occasionally exclude an insignificant number of items.

This report includes the available data as of May 2007; several levels, either where the spatial exposure is minimal (i.e., Layer VI-14), or where no archaeological material was observed during fieldwork (i.e., Layers II-2, II-3), are not presented in this work.

Included in this report are 15 archaeological layers (Table 2.1); the analysis of each of these layers incorporates the following components:

Table 2.1 The archaeological layers included in the study, from the topmost layer of the stratigraphic sequence (younger) to the lowermost (older); area in m^2, volume in m^3

		Layer	Level	Area[a]	Volume[b]	Lithics counts	
						>2 cm*[c]	≤2 cm*[d]
GBY excavation areas	C	V-5		6.39	1.59	408	36,770
		V-6		7.04	1.97	356	6,585
	A	I-4		5.25	1.57	32	6,696
		I-5		5	0.55	63	15,350
	B	II-2/3		4.67	0.47	139	7,502
		II-5		25	13	180	3,903
		II-5/6		19.14	0.38	142	10,531
		II-6	L-1	23.79	4.28	2,295	58,086
		II-6	L-2	25.62	3.07	1,412	79,670
		II-6	L-3	17.92	2.50	1,199	96,094
		II-6	L-4	16.64	2.16	1,729	118,434
		II-6	L-4b	13.69	0.82	768	8,778
		II-6	L-5	13.39	1.20	450	37,609
		II-6	L-6	12.62	1.38	732	13,357
		II-6	L-7 upper occupational horizon	12.60	1.38	1,098	25,915
		II-6	L-7 northern test pit	2.75	1.51	332	12,555
		II-6	L-7 southern test pit	4.25	2.89	104	6,874
		Total		215.76	40.72	11,439	544,709

[a]Area represents the spatial extent of the excavated material (see Section 2.4.5)

[b]Volume is the excavated area multiplied by the estimated mean of excavated thickness based on cross-sections

*Lithic counts represents the total number of lithic artifacts of all raw materials including: [c] items larger than 2 cm (i.e., macroartifacts: flakes and flake-tools, cores and core-tools, and bifacial tools and [d] smaller items (i.e., microartifacts)

[4]Layers I-4 and II-5, revealed during the first season of excavations at the site, were partially wet-sieved; sampling included a single full bucket from each sub-square of a given depth unit of excavation.

1. All the lithic material retrieved in the course of excavation (flint, basalt, and limestone); including microartifacts, macroartifacts, and pebbles (described below under "lithic analyses").
2. Lithic material retrieved through sorting of the wet-sieved sediments:
   - Microartifacts, including flint, basalt, and limestone
   - Cores and core-tools (items lacking a ventral face)
   - Pebbles

Excluded from this analysis are the flakes and flake-tools retrieved from the sorting of the wet-sieved sediments and natural small stone items that do not bear signs of knapping and hence are not identified as microartifacts (these are discussed in detail below, under "Lithic Analyses").

2.4.3 *Lithic Analyses*

The lithic assemblages referred to in this study comprise various raw materials, including flint, basalt, and limestone. In addition, they consist of both natural items (i.e., pebbles) and modified items – macroartifacts (>2 cm) and microartifacts (≤2 cm). The analyses of the lithic assemblages are specified below, according to the different lithic categories.

2.4.3.1 Microartifacts

The main component of this study is the numerous microartifacts (≤2 cm), retrieved from the various archaeological layers, during excavation or from the sorting of the wet-sieved sediments. Unlike macroartifacts, flint microartifacts occur in extremely large quantities in each of the archaeological layers, thereby providing large enough samples of burned items for spatial analysis.

A large number of stone microartifacts was retrieved from the different archaeological occurrences; analyses of these items were carried out in the following steps:

1. *Differentiating between natural and modified items*: in order to ensure that the examined spatial patterns represent evidence of hominin activity, only items that are unquestionably the result of stone knapping are included in this study. Thus, natural items (e.g., small pebbles) are excluded. Only items that exhibit characteristic knapping features of flaked material (e.g., ventral face, striking platform) are included in this category and are defined as microartifacts.
2. *Defining raw material*: microartifacts are sorted into different classes of raw material, including basalt, limestone, and flint. The general inventory of these is presented for each of the archaeological layers.
3. *Identifying burning damage on flints*: the final and most vital stage of analysis is the identification of burning damage on the flint microartifacts (equivalent principles are used for the identification of burning on flint macroartifacts and flint pebbles). Identification is based on the presence of typical macrofractures (i.e., potlid fractures), known to result from the exposure of flint to high temperatures (see the detailed description above under "Research Hypotheses"). The identification of burning is thus based entirely on visual observation (Appendix 1).

During this research, samples of burned flint microartifacts were submitted to thermoluminescence (TL) analyses.[5] The results of the TL study have demonstrated that the analyzed samples must have been exposed to high temperatures in a heating event in the remote past (Alperson-Afil et al. 2007); these results provide independent verification that the observed potlid fractures are indeed the result of burning.

2.4.3.2 Macroartifacts

This category comprises artifacts larger than 2 cm of three different types: cores and core-tools (i.e., CCT), flakes and flake-tools (i.e., FFT), and bifacial tools (i.e., handaxes and cleavers). The general inventory of these is presented for each of the studied archaeological layers by raw material; spatial distribution is examined only for the burned flint macroartifacts.

2.4.3.3 Pebbles

Of the various analyzed lithic components of this study, unmodified items are included only within this category. The category classified here as "pebbles" refers to natural items larger than 2 cm (i.e., pebbles and cobbles) and consists of flint, basalt, and limestone. The general inventory of the pebble assemblage is reported for each of the analyzed archaeological layers; spatial distribution is examined only for burned flint pebbles.

2.4.4 *Database Construction*

The analyzed lithic assemblages are organized in two different types of *Access* database. This distinction originates in the use of two different databases of lithic analyses in the excavation project of GBY. In the first, each database

[5] The principles of the TL method are discussed in Section 1.2.

row incorporates the attributes of a single item, whether a macroartifact or a microartifact. The second type of database consists only of microartifacts retrieved through sorting of the sieved sediments; here, each row incorporates the total content of a sediment unit (i.e., of an excavated spit). The difference between these two databases required separate procedures in order to convert the data into spatially manageable geographical information; more specifically, the database in which the entire content of an excavated spit was depicted in a single row had to be converted into a "single-record row" database. This conversion enabled spatial plotting of all items, discussed below.

For each of the lithic items, the recorded data includes: stratigraphic assignment, provenance recording (either a full X, Y, Z reference or a 0.5 × 0.5 m quadrant and a range of elevations), raw material definition, and in the cases of flint items the presence or absence of burning damage.

2.4.4.1 Assigning Artificial Coordinates

A large number of macroartifacts and the majority of microartifacts were retrieved with a general spatial reference, either during excavation or throughout the sorting of the wet-sieved sediments. The spatial reference of these includes the X and Y quadrant (0.5 × 0.5 m) and depth of spit (Z is a range of depths). Such spatial recording allows only the representation of relative frequencies of lithic items per excavated unit. Other spatial analyses, such as creating a density map, would necessitate measuring the distances between different features and thus require that the data be depicted as distinct points.

It has been suggested that assigning a random spatial reference within the excavated area provides a reliable, and almost identical, spatial representation (Gilead 2002). Taking this into consideration, using the *Visual Basic* language within the *Access* program (Microsoft®Accesss 2002) items with a general spatial reference were given a new reference point within their recorded sub-square. This procedure enabled the plotting of each of the excavated lithic finds and included the following stages:

Each of the archaeological layers was treated independently within a separate database. The database was then sorted according to the recorded excavated units of the particular layer. Each of these excavated units had a defined excavated area (0.5 × 0.5 m sub-squares or 1 × 1 m squares), from which a certain number of lithic items was retrieved. This area (a) was then divided by the maximum value of items retrieved from that area (n) so that each item could be plotted separately within an a/n area (δ). Let us hypothesize a case in which a given 1 m^2 excavated area ($a = 1$) has 100 flint items ($n = 100$). If these 100 items were distributed *evenly* within the 1 m^2 area each item would occupy an area of 1/100 m^2 ($\delta = 0.01$). The new reference point for each of these items is defined as the southwestern corner of each δ cell, so that:

$$a = \sum \delta_{1-n} \left(\delta_1 + \delta_2 + \delta_3 + \delta_4 \cdots \delta_n\right)$$

This procedure enables the items to be plotted *uniformly* within their recorded spit, ensuring that the new plotted data are as consistent as possible with the recorded data of sub-square precision. Other plotting methods (e.g., random plotting) may have resulted in the formation of artificial clusters within the area of the sub-square.

In addition, it is important to note that the analysis of spatial patterns in this study is carried out on the data according to their original sub-square recording (see Sections 2.4.6.1 and 2.4.6.3) and that the point-plotted data are used mainly for illustrating the observed patterns of distribution (i.e., in the density maps of kernel type – see Section 2.4.5).

Several procedures required a three-dimensional representation of the data (e.g., assigning a stratigraphic classification, specified below). In these cases the vertical position (i.e., Z coordinate) of the items was essential. As previously discussed, many "uncoordinated" pieces were retrieved from the field with a recorded *range* of elevations. Due to the tilted position of the archaeological exposures, elevations were recorded in two corners of the excavated unit, northeastern (NE) and southwestern (SW), at the beginning (TOP) and end (BOTTOM) of each excavation phase (defined as a 5 cm spit of excavated material). In order to convert these elevations into a single Z point, the average of the recorded elevations was calculated so that the new Z point represents the elevation at the center (both vertical and horizontal) of the excavated unit:

$$NEW\,Z = \frac{\left\{\left[(NETOP + NEBOTTOM)/2\right] + \left[(SWTOP + SWBOTTOM)/2\right]\right\}}{2}$$

This procedure enabled analysis of a small number of layers that required additional treatment in order to allow spatial plotting of the excavated material:

Layers I-4 and I-5: these two layers were exposed during 1989, the first season of renewed excavations, when fieldwork focused on two areas; the southeastern part of the study area (Area A) and some 45 m to the northwest (Area B) (Fig. 1.3).

In Area A, the tilted nature of the archeological occurrences was revealed during excavation. Upon the quarrying of Trench I, each of the observed archaeological layers was assigned an individual reference name (i.e., I-4 and I-5). These two layers, observed in various sections within the excavated area, exhibited a sedimentological divergence between gray clay (I-4) and a coquina mixed with sandy and clayey lenses (I-5). As excavations proceeded, the distinction between these two horizons became evident and material was given a definite stratigraphical assignment. However, for some of the excavated material a stratigraphic assignment was not specified. These circumstances resulted in an excavated assemblage in which some of the material is recorded with a full spatial reference (i.e., excavated grid unit, range of elevations, and specific layer), while other material lacks registration of the stratigraphic assignment.

In order to allow spatial plotting of the excavated material from Area A, it was necessary to determine the stratigraphic position of some of the excavated assemblages. Using *ArcScene* (ESRI®ArcScene™9.3), the three-dimensional data analysis software available in the ArcGIS package,[6] the entire assemblage of Area A was plotted three-dimensionally and then divided into two separate stratigraphical units. The division was enabled through the use of a "virtual" 3D surface, designed to depict the tilted contact between I-4 and I-5. The outlines of this "contact surface" follow the contact lines of I-4/5 as drawn in the various field cross-sections; thus items above the surface were assigned to I-4 and items below it to I-5.

In addition, in order to enlarge stratigraphic clarity during fieldwork in Area A, the area was excavated on either side of a baulk (Fig. 1.3); thus the spatial exposure of these layers is not continuous. Furthermore, during that season, the excavation of the relatively sparse exposure and density of Area A came to an end before more extensive exposure of the layers, and fieldwork then focused on the denser occurrences of Area B.

Accordingly, in this study the data presented for Area A include the general lithic inventory of the entire exposed surfaces, while the spatial presentation involves only one area, to the north of the baulk, where the excavation reached Layer I-5; the spatial account is thus minimal and no further spatial analyses are carried out for these layers.

Layer II-6 L-7: this stratigraphic unit is the lowermost occupational level of Area B, excavated during the 1995–1997 seasons. The upper part of Layer II-6 L-7 revealed an occupational surface embedded within a sandy matrix made up primarily of crushed mollusks.

At the end of the 1996 season, excavations in two areas (one in the southern part and another at the northernmost edge of the exposed surface) completed the exposure of the upper occupational horizon of Layer II-6 L-7 and items were drawn and removed from the excavated surface. Thus, during the 1997 season excavation in Layer II-6 L-7 penetrated deeply into two test pits at the edges of the excavated surface, while excavation of the central part of the surface uncovered the upper occupational horizon. Excavation of these two test pits reached the bottom of Layer II-6 L-7 – the contact between Layer II-6 and the underlying Layer II-7. These two test pits revealed a sorted sedimentological sequence, with a very coarse conglomerate at the base that fines upwards, exhibiting a thin clayey layer above the conglomerate and a thick series of sands above it.

The entire layer, from the conglomeratic base (with its two test pits) to the sandy top, was designated II-6 L-7. Due to these sedimentological differences, items retrieved from the two test pits had to be separated from the general assemblage of the upper occupational horizon of Layer II-6 L-7. This was accomplished by isolating the material retrieved from the specific excavated units of the test pits during the 1997 season (as excavation season is specified in the databases). The upper occupational horizon of Layer II-6 L-7 thus includes material from these two areas excavated during the 1995–1996 seasons as well as the entire exposed surface of Layer II-6 L-7 from the central area between the test pits.

As in Area A, the discontinuous nature of these test pits does not allow spatial analysis; thus the data displayed for the Layer II-6 L-7 test pits consist of the general lithic inventory, and only a schematic spatial illustration of the flint microartifacts is presented. A similarly brief description is given for Layer II-2/3, which was exposed over a relatively small area (4.67 m^2) that does not permit in-depth spatial analysis.

2.4.5 Generating Distribution Maps and Density Maps

The assignment of artificial coordinates to the lithic microartifacts and macroartifacts enabled the various databases of lithic material to be used as geographical information that can be integrated into ArcGIS software. This package is a collection of software and geographic data for capturing, managing, analyzing, and displaying all forms of geographically referenced information. Of the software available, *ArcMap* (ESRI®ArcMap™9.3) was used for the spatial display and analyses of the archaeological data in this study.

An individual *ArcMap* project was designed for each of the analyzed archaeological layers. The databases of lithic items were then inserted into the *ArcMap* file, each depicted as a separate layer of geographical information.

Following insertion of the data, a systematic methodology of spatial display and analysis was maintained (see Appendix 2).

The initial phase of analyses consisted of evaluating the spatial distribution of the point-plotted items, which can be illustrated in regular point-distribution maps (Fig. 2.1a). These illustrations are used in this study to display the distribution of macroartifacts. However, when the lithic

[6] See: http://www.esri.com/software/arcgis/about/desktop_gis.html

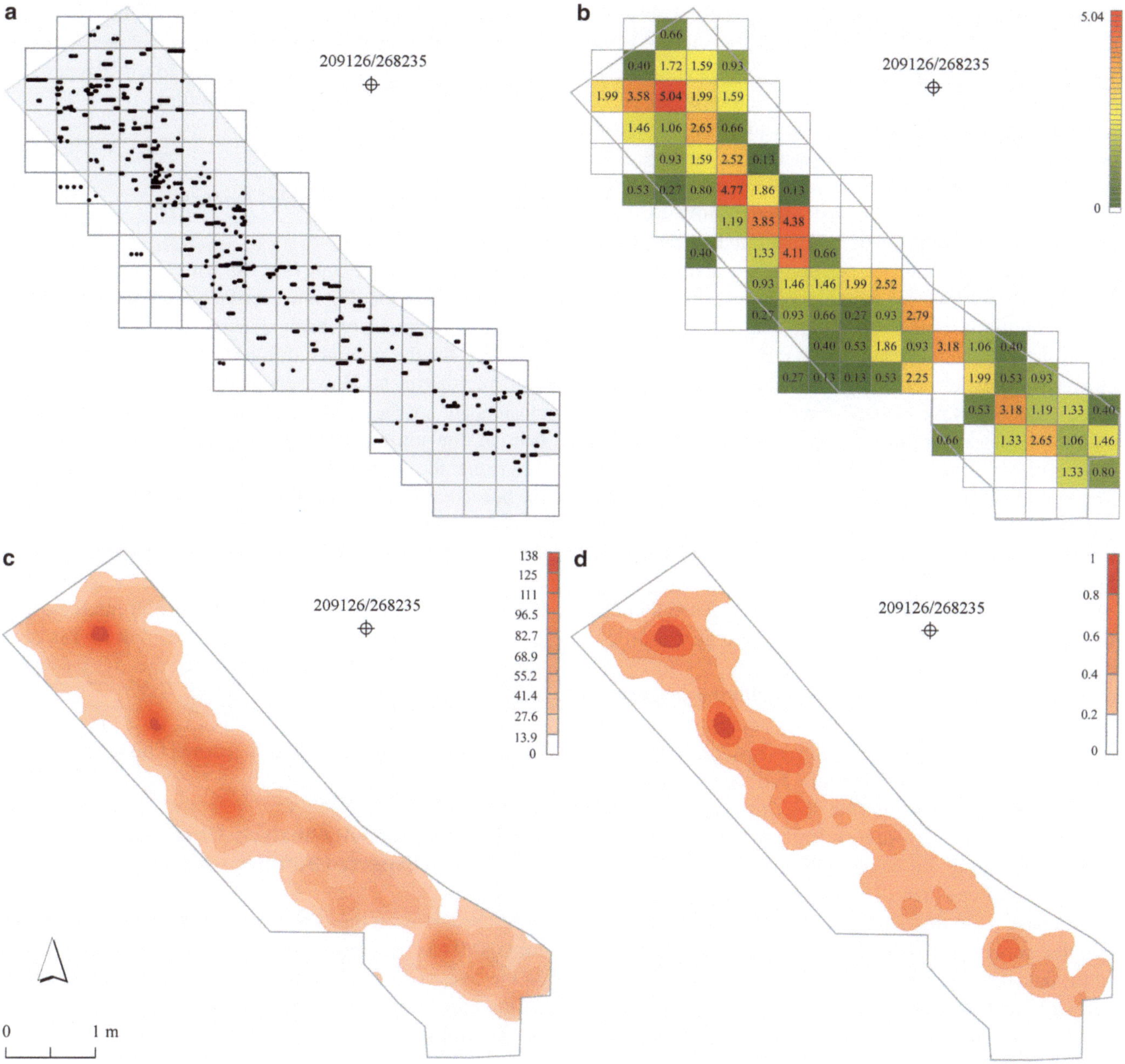

Fig. 2.1 The stages of building distribution and density maps, demonstrated on the assemblage of burned flint microartifacts from Layer II-6 L-1 (N = 754): (**a**) point-plotted distribution map; (**b**) percentages of microartifacts per excavated unit; (**c**) kernel density map; and (**d**) standardized kernel density map

category contains extremely large quantities, as is the case for microartifacts, it is impossible to distinguish areas of high density within the general distribution pattern. This necessitated a schematic illustration of the relative percentages of microartifacts per excavated unit. In order to emphasize areas of high density, these illustrations use a green-to-red color scheme, which depicts different degrees of frequency by gradually altering colors (green for low frequencies, red for high frequencies) (Fig. 2.1b).

In order to achieve a reliable representation of the extent of the excavated area of each layer, we have depicted the margins of the excavated area according to the distribution of the archeological finds that is available from the field maps of the living floors (available for Layers V-5, V-6, and II-6 L-1 to L-7) and from the distribution of coordinated pieces (i.e., items that were given full XY coordinates in the field; these include FFT, CCT, bifacial tools, and pebbles). The spatial extent of these items is the optimal representation of the boundaries of the excavated area of each archaeological layer. The outline of this area is illustrated in all distribution and density maps. As some of the excavated material, particularly the microartifacts, was given artificial coordinates within the excavated spit (i.e., square or sub-square), the graphic representations of percentages of microartifacts per excavated unit, which refer mechanically to grid sub-squares, include areas that are not fully excavated (e.g., a full sub-square will be illustrated in the graphic presentation of percentages per excavated unit even where only the corner of the sub-square was excavated).

In order to illustrate areas of high density graphically, the point-plotted data of microartifacts distribution were converted into kernel density maps (Fig. 2.1c). Kernel density calculates the density of point features around each output cell (determined in this study as 0.01 m). Conceptually, a smoothly curved surface is fitted over each point. The surface value is highest at the location of the point, and diminishes with increasing distance from the point, reaching 0 at the search radius distance from the point, determined here as 0.5 m; thus, only a circular neighborhood is possible. The density value at each output cell is calculated by adding the values of all the kernel surfaces where they overlie the cell center.[7]

Determination of different search radii thus changes the scale of the analysis results. With a smaller radius, fewer points will fall within the search radius, resulting in numerous small, "dense" features. Increasing the radius will result in more points falling within the search radius; this number (of points) will be divided by a larger area when calculating density, resulting in larger, generalized concentrations. The values of cell size (0.01 m) and search radius (0.5 m) were chosen for this study as they closely represent the genuine patterns observed within the schematic illustrations of the data (see Appendix 3, where a comparison of different cell sizes and search radii is illustrated in comparison with the density patterns as depicted through data interpolation of sub-square precision [methodological procedures are specified in Appendix 2]). Finally, in order to create a uniform scale (from 0 to 1) that will enable comparison between kernel density maps of different data sets (e.g., in-between layers; burned vs. unburned flint), the densities have been standardized by the maximum values of each data set (Fig. 2.1d).

In this study, kernel density maps are produced only for microartifacts, since these occur in large numbers that do not allow evaluation of spatial patterns in their "point-plotted" form. A uniform scale (from 0 to 1) with five levels of density is applied to all the density maps (the lowest density level is 0–0.2 and the highest is 0.8–1.0). In addition, a uniform color scheme is applied to the kernel density maps, in which each lithic category is depicted in a different color; blue for unburned flint, red for burned flint.

2.4.6 *Analysis of Spatial Patterns*

The initial stage of analysis, in which distribution and density maps of the flint microartifacts were produced, drew attention to areas of high density and provided basic evidence for the presence or absence of clusters of burned flint microartifacts. However, in order to verify that these clusters are not the random outcome of the original distribution of the entire flint component, it was essential to determine the degree of overlap between the distribution of the burned and unburned flint microartifacts.

As thoroughly discussed above (see Section 2.3), when the burned and unburned flint microartifacts overlap absolutely we cannot rule out the possibility of a natural fire. Conversely, when the clusters of burned flint microartifacts do not coincide with those of the unburned flint we can plausibly suggest that an anthropogenic fire is the agent responsible.

Several methods were applied to examine the degree of overlap between the burned and unburned flint microartifacts.

2.4.6.1 Homogeneity Analysis: Observed and Expected Burning

This method examines the distribution of the burned flint microartifacts in comparison with that of the unburned ones. In the case of an absolute overlap between the distributions of the burned and unburned flint microartifacts, we expect the relative percentage of burned items to be homogeneous across the exposed surface, displaying similar values in each of the excavated grid units. Thus, if the general percentage of burned flint microartifacts in a particular layer is 2.00%, we expect that within each of the excavated units (i.e., 0.5 × 0.5 m sub-squares) the percentage of burned items within the total flint microartifacts of the sub-square will similarly be 2.00%.

In order to compare between the observed and expected percentages of burned items in each excavated unit, the expected percentage of burned flint microartifacts was subtracted from the observed percentage. The value obtained through this calculation is the deviation between the observed and expected percentage of burning in each excavated unit; units of positive values are excavated sub-squares in which the observed percentage of burning exceeds the one expected in the case of uniform distribution of the burned flint microartifacts (see detailed procedures in Appendix 2).

[7] The kernel function is based on the quadratic kernel function described in Silverman 1986: 76, Equation 4.5.

2.4.6.2 Generating Random Patterns

This method of generating random patterns attempts to illustrate patterns of density in a case of random distribution of burning across the exposed surface. Three independent random scenarios were sequentially produced for each of the analyzed archaeological layers (see detailed procedures in Appendix 2). A "random selection" tool was used in order to produce a random selection of a particular number of items out of the entire assemblage of flint microartifacts. The number of randomly selected items is equivalent to the number of burned flint microartifacts recorded in the analyzed layer. This procedure was sequentially repeated three times. Next, for each data set of randomly selected flint microartifacts a kernel density map was produced, following the same criteria (i.e., cell size, search radius, and scale normalization) as in the other kernel density maps of this study (detailed above). A green color scheme is consistently used for the random density maps.

These procedures yielded three possible scenarios of random densities. Discrepancies between these and the observed density patterns of the burned flint microartifacts can be used as an additional indicator of the significance of the observed patterning of burned flint microartifacts.

2.4.6.3 Statistical Tests

The ArcGIS package supports various types of spatial statistic tools (e.g., cluster analyses, nearest-neighbor analysis, etc.) However, as discussed previously (see Section 2.3.3), differentiating the patterning of the burned flints from the unburned ones is not a straightforward issue. The burned flint microartifacts spatially originate from the larger flint component (in each analyzed layer), which may *a priori* be spatially clustered; thus we cannot consider the burned flint microartifacts a spatially distinct sample on which spatial statistic analyses can be performed. If we did this, we would have failed to notice the possible overlapping of the burned and unburned flints, which is a fundamental factor in a reliable identification of anthropogenic fire.

A chi square test, however, can examine the spatial differences between the burned and unburned flint microartifacts, providing a statistical parameter of probability for that differentiation. The chi square (χ^2) value was thus calculated for the burned flint microartifacts over all the excavated units (i) through the following equation:

$$\chi^2 = \sum_i \frac{(OBS_i - EXP_i)^2}{EXP_i}$$

so that the absolute chi square test value of a particular archaeological layer is the summary of χ^2 values of all excavated units (i = number of excavated units).

The probability level (p) of the chi square test is then extracted by comparing the calculated chi square value to a critical value from a chi square table, with degrees of freedom corresponding to that of the data ($df = i - 1$).

The chi square goodness of fit supplies a parameter of differentiation between the observed distribution and an expected, uniform, distribution. It does not indicate, however, what specifically is significant. This can be portrayed in the standardized residuals (SR), which are the signed square root of each category's contribution to the χ^2:

$$SR = \frac{OBS_i - EXP_i}{\sqrt{EXP_i}} \sim N(0,1)$$

What the above formula actually states is that the standardized deviations are approximately (asymptotically) normally distributed. i.e., given a large enough sample and a sufficient number of units, one would expect (under the assumptions of the null hypothesis) that about two thirds of the units will have SR values in the −1 to +1 range, about 95% will be between −2 and +2, etc. Thus, any unit for which the SR value is greater than 2.00 (and the expected value is larger than 5) is considered a substantial contributor to the significance observed in the chi square test (e.g., Haberman 1973).

Standardized residuals were thus calculated for the burned flint microartifacts of each excavated unit. Where burned flint microartifacts are distributed significantly different from the unburned ones, we can evaluate the contribution of different excavated areas to the observed difference.

2.4.6.4 Analysis of High-Density Clusters

The previous sections have outlined the methodologies for the identification of significant clusters of burned flint microartifacts. These clusters are the ones that display high levels of density (i.e., reaching the fifth and highest recorded density level), and are distributed significantly different from the unburned flint microartifacts. Such clusters are interpreted here as possible remnants of anthropogenic fires (i.e., hearths). Following their identification, these clusters are examined with reference to the distribution of other burned flint items (i.e., FFT, CCT, and pebbles).

The attempt to characterize the clusters of burning is further accompanied by various measurements of the kernel of the clusters, where the highest level of density is recorded. The measurements refer to the general geometry (area and diameter) and lithic composition (relative percentages of burned and unburned flint microartifacts).

Chapter 3 presents the results obtained through the use of the various methodologies on some 15 archaeological occurrences within the GBY depositional sequence.

Chapter 3
Results

This chapter integrates the data on the spatial distribution of burned flint items and the general lithic inventory of the different occupation levels in Gesher Benot Ya'aqov. The analyzed assemblages are presented in chronological order, from the topmost (i.e., youngest) layer of the stratigraphic sequence to the lowermost (oldest) one.

First, the lithic composition of each layer is presented in a summary table specifying different lithic categories; i.e., microartifacts, flakes and flake-tools (FFT), cores and core-tools (CCT), bifaces (handaxes and cleavers), and pebbles. These are summarized according to the different raw materials used (i.e., flint, basalt, and limestone). The percentages of burned flint items within the flint component of each lithic category are incorporated in these summary tables.

Following that, the spatial data for the flint items are presented. As the main objective of this study is to examine the spatial distribution of the burned flint items to detect possible concentrations of burning, only data for the flint microartifacts and macroartifacts are presented.

3.1 Layer V-5

The lithic assemblage of Layer V-5 consists of 36,770 microartifacts and 408 macroartifacts of various raw materials, retrieved from an area of 6.39 m^2 (Table 3.1). Flint is the dominant raw material, displaying evidence of burning on 1.81% of the microartifacts and 0.31% of the macroartifacts (Table 3.1).

Burned flint microartifacts are distributed throughout the exposed surface and are recorded in 28 of the 35 excavated sub-squares (Fig. 3.1). Three areas of relatively high frequencies are observed: in the northern part, 15.71% of the burned flint microartifacts are concentrated within an area of 0.5 m^2 (i.e., two sub-squares); in the central part, three sub-squares incorporate an additional 19.11%; the highest percentages of burned flint microartifacts (22.86%) are recorded within two sub-squares in the southern part of the excavated area (Fig. 3.1).

For unburned flint microartifacts, the highest percentages are likewise observed in the southern part of the exposed surface. Here, however, high percentages of microartifacts, adding up to 33.98% of the unburned flint microartifacts of the layer, occur within a larger area (1 m^2) (Fig. 3.1). The three sub-squares of the central area, in which 19.11% of the burned flint microartifacts are recorded, also display relatively high percentages of unburned flint microartifacts (13.70%). In the northern area, where 15.71% of the burned flint microartifacts are recorded within two sub-squares, high percentages of unburned flint microartifacts are observed in a single sub-square (6.72%).

The distributions of the burned and unburned flint microartifacts thus generally overlap each other; this is further illustrated in the density maps, both showing a single high-density concentration in the southern area of the exposed surface (Fig. 3.2). The concentration of the unburned flint microartifacts is, however, larger and more diffuse than that of the burned ones (Fig. 3.2). A similar configuration is illustrated in the random density maps, which all point to the southern area as the area most likely to exhibit a high-density concentration of burning (Fig. 3.3).

Despite the general overlapping of the burned and unburned flint microartifacts, the chi square test for the burned flint microartifacts suggests that their distribution is significantly different from an expected, uniform distribution ($\Sigma\chi^2 = 176.43$; $df = 33$; $p < 0.001$). Calculation of standardized residuals for burned flint microartifacts of different excavated units suggests that the areas that contribute to this pattern (SR > 2) lie mostly outside the southern congruent concentration, in sub-squares in which the observed percentage of burning exceeds the expected one. Included in these are the two northern sub-squares that display a gap of 3–4% between the observed and expected percentages of burning (Fig. 3.4).

In the western one the observed percentage of burned flint microartifacts is 3.22% higher than the expected one (SR = 2.93; N [expected] = 37.92), and in the eastern one the observed percentage of burning is 3.68% higher than the expected one (SR = 6.12; N [expected] = 11.34).

Slightly to the south, in the central area, three additional sub-squares exhibit similar patterns: in the northern of these the observed percentage is 2.80% higher (SR = 3.78; N [expected] = 17.25), in the southern one the observed percentage is 2.41% higher (SR = 2.45; N [expected] = 30.44), and in the eastern of the three the observed percentage is 2.04% higher (SR = 6.04; N [expected] = 3.57).

Near the southern concentration of burned flint microartifacts, only one sub-square (which covers the northwestern edge of the concentration) exhibits a relatively significant SR value (SR = 2.23; N [expected] = 3.08); the observed percentage of burned flint microartifacts within this sub-square is only 0.69% higher than the expected one.

In Layer V-5, burned flint microartifacts are not distributed evenly and display various areas of high frequencies throughout the excavated area. The density map illustrates a single high-density concentration in the southern part of the exposed surface (Fig. 3.2). This concentration does not coincide entirely with that of the unburned flint microartifacts and its high-density kernel (covering an area of 0.116 m^2) includes 9.46% of the burned flint microartifacts of the layer and only 4.06% of the unburned ones.

Out of the flint macroartifacts of Layer V-5, only a single burned item was observed (Table 3.1). This item was recorded

Table 3.1 Lithic assemblage of Layer V-5

Category	Flint				Basalt	Lime-stone	Total
	Unburned		Burned				
	N	%	N	%	N	N	N
Microartifacts[a]	30,315	98.18	560	1.81	5,562	333	36,770
FFT artifacts[a]	289	100.00	–		74	1	364
CCT artifacts[a]	31	96.87	1	3.12	8	3	43
Handaxes	–		–		–	–	–
Cleavers	–		–		1	–	1
Pebbles	69	100.00	–		165	24	258
Total	30,704	98.20	561	1.79	5,810	361	37,436

[a] The percentage of burned and unburned flint items is calculated within each lithic category

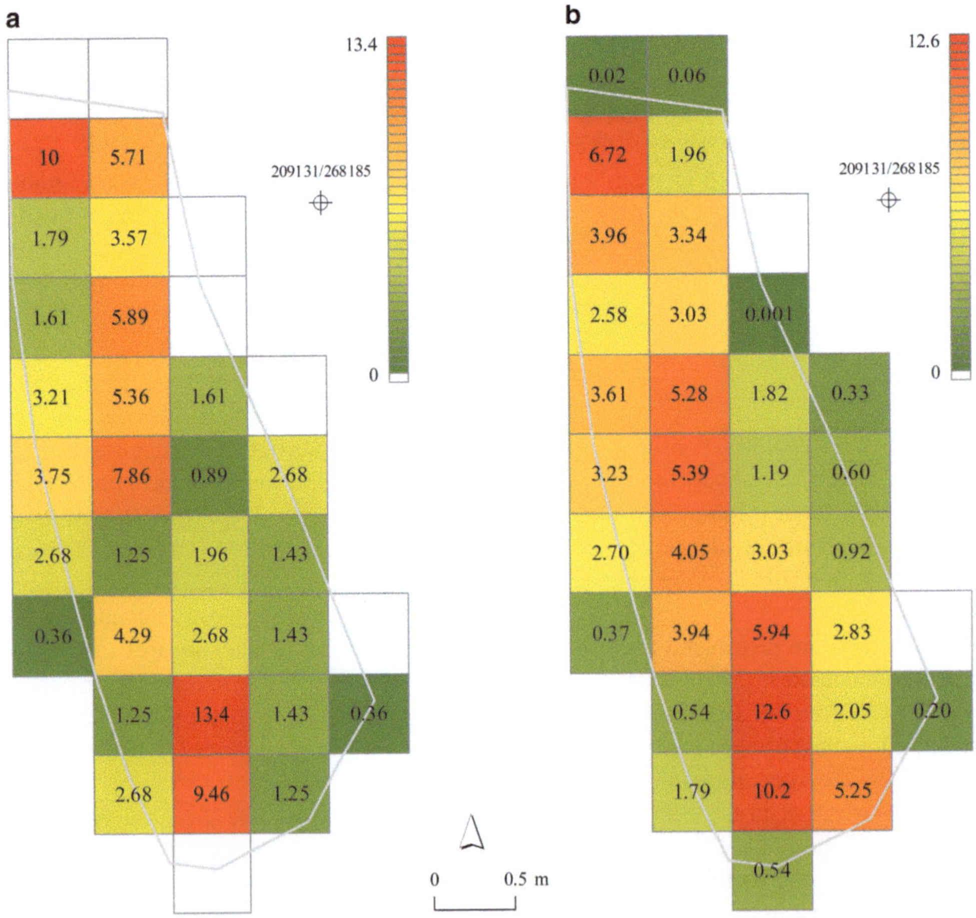

Fig. 3.1 Percentages of flint microartifacts per excavated unit in Layer V-5. (**a**) Burned flint microartifacts (N = 560) and (**b**) unburned flint microartifacts (NV = 30,315)

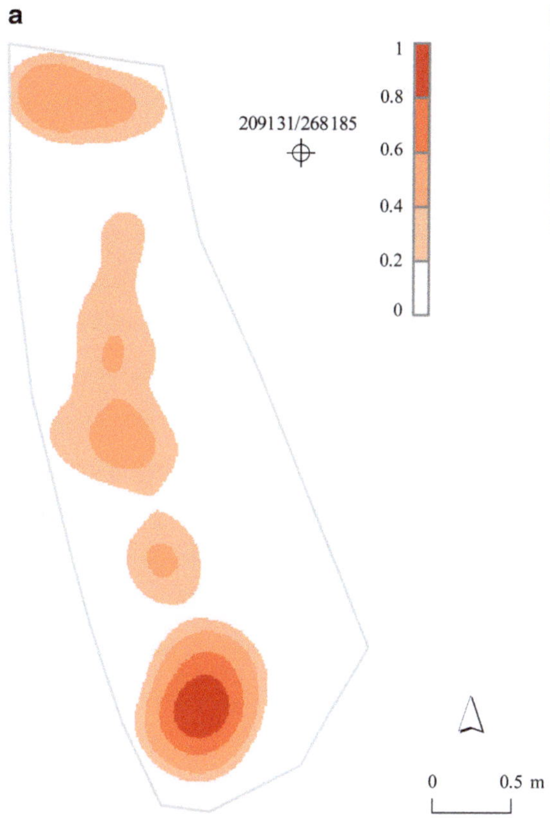

Fig. 3.2 Kernel density maps of flint microartifacts in Layer V-5. (**a**) Burned flint microartifacts (N = 560) and (**b**) unburned flint microartifacts (N = 30,315)

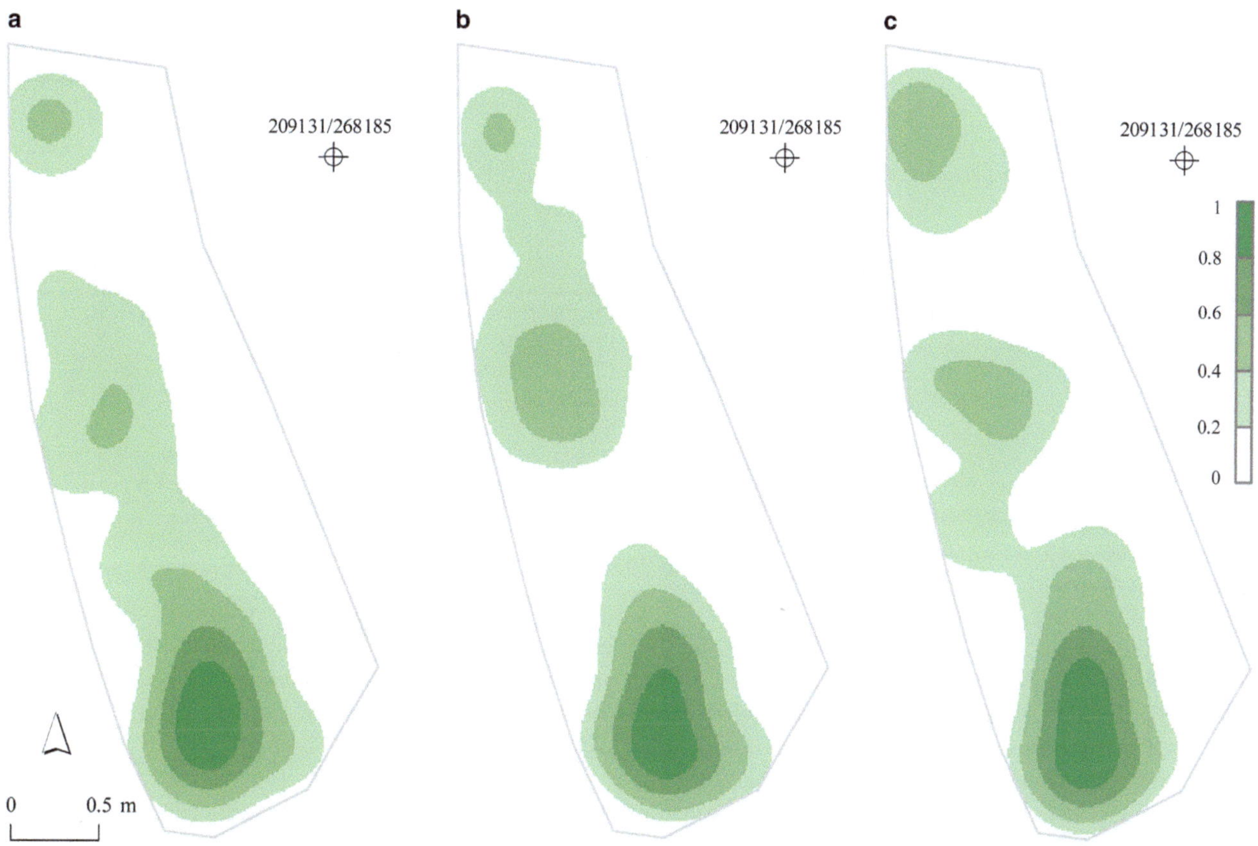

Fig. 3.3 Kernel density maps of three randomly selected data sets (N = 560) for Layer V-5

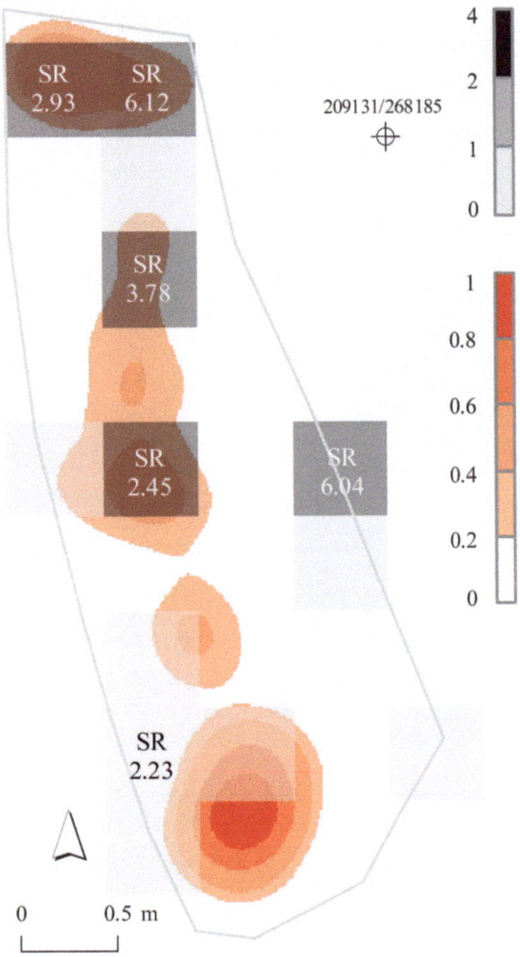

Fig. 3.4 The kernel density map of burned flint microartifacts of Layer V-5 and excavated units in which the observed percentage of burning exceeds the expected percentage; significant SR values are marked on the map

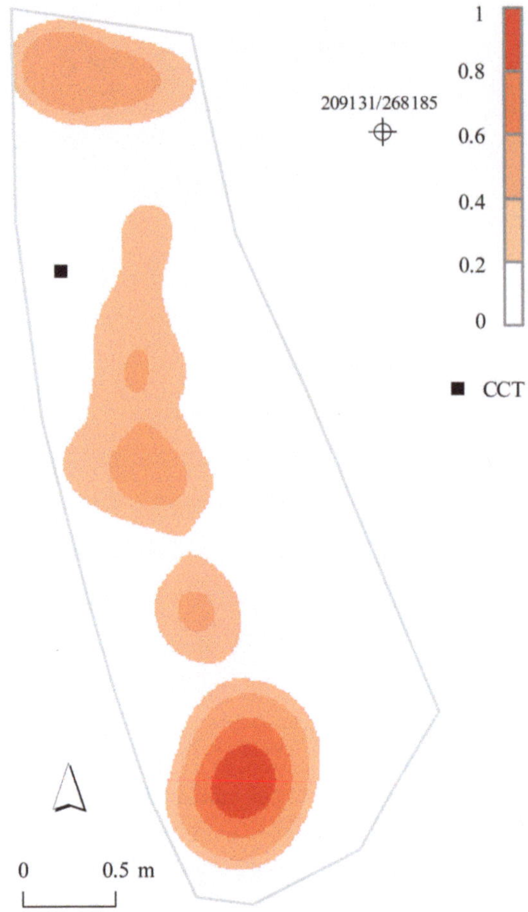

Fig. 3.5 The kernel density map of burned flint microartifacts of Layer V-5 and the distribution of large burned flint items (CCT: N = 1)

some 3 m to the northwest of the high-density concentration of burned flint microartifacts (Fig. 3.5).

3.2 Layer V-6

The 7.04 m^2 of exposed surface of Layer V-6 yielded 6,585 microartifacts and 356 macroartifacts of various raw materials, modified predominantly on flint (Table 3.2). Evidence of burning was recorded on 1.84% of the flint microartifacts and 0.70% of the flint macroartifacts (Table 3.2).

Table 3.2 Lithic assemblage of Layer V-6

Category	Flint				Basalt	Lime-stone	Total
	Unburned		Burned				
	N	%	N	%	N	N	N
Microartifacts[a]	4,424	98.15	83	1.84	1,982	96	6,585
FFT artifacts[a]	273	99.27	2	0.72	46	1	322
CCT artifacts[a]	9	100.00	–		19	–	28
Handaxes	1		–		–	–	1
Cleavers	–		–		5	–	5
Pebbles	5	100.00	–		49	3	57
Total	4,712	98.22	85	1.77	2,101	100	6,998

[a]The percentage of burned and unburned flint items is calculated within each lithic category

Burned flint microartifacts are recorded in 18 of the 37 excavated sub-squares. The burned flint microartifacts occur in varying frequencies within the different excavated units (Fig. 3.6). The highest percentages are observed in the central part of the excavated area; there, 20.50% of the burned flint microartifacts occur within a single sub-square, and the two sub-squares to its north together incorporate an additional 14.46% (Fig. 3.6). Thus, 34.96% of the burned flint microartifacts are recorded within an area of 0.75 m^2. High percentages of burning are also observed in three sub-squares in the eastern part of the exposed surface (comprising 19.27%) and in two sub-squares in the northwestern corner (13.25%) (Fig. 3.6).

Unburned flint microartifacts extend throughout the excavated surface, covering a larger area (Fig. 3.6). Relatively high percentages are recorded in the central part of the exposed surface; 34.23% of the unburned flint microartifacts

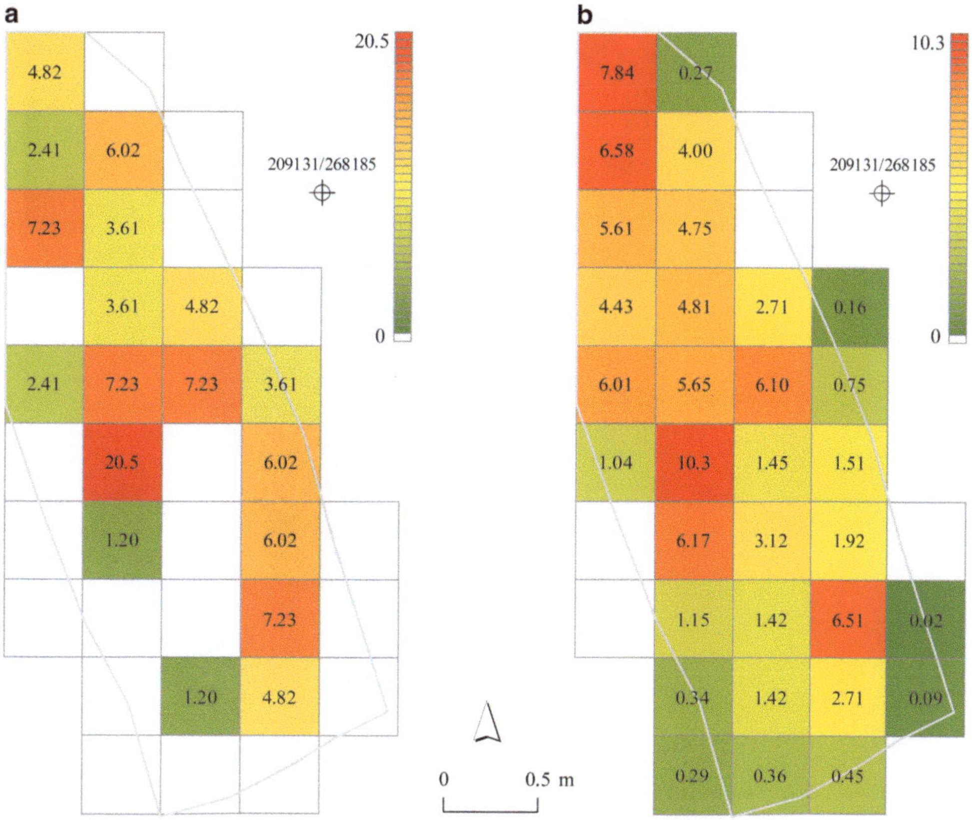

Fig. 3.6 Percentages of flint microartifacts per excavated unit in Layer V-6. (**a**) Burned flint microartifacts (N = 83) and (**b**) unburned flint microartifacts (N = 4,424)

are included within five sub-squares, one of these yielding 10.30% (it is within this sub-square that 20.5% of the burned flint microartifacts occur). An additional 20.03% is included within three sub-squares in the northwestern corner of the surface (Fig. 3.6).

The burned and unburned flint microartifacts display comparable patterns of distribution; however, the unburned ones are present in most of the excavated units and their zones of high frequencies seem to extend over larger areas.

Density maps illustrate the large degree to which the concentrations of the unburned flint microartifacts extend in comparison with the smaller dense concentrations of the burned ones (Fig. 3.7). Two high-density concentrations are illustrated in the density map of the burned flint microartifacts. While the larger (southern) one partly overlaps the high-density kernel of the unburned flint microartifacts' concentration, the other (northern) one does not (Fig. 3.7).

This northern concentration of burned flint microartifacts is not depicted in any of the three random density maps (Fig. 3.8). The southern concentration is illustrated in one of the three random density maps, where, however, it does not exhibit the highest density level (Fig. 3.8).

Applying a chi square test to the burned flint microartifacts suggests that the difference between the observed distribution and an expected, uniform one is not highly significant ($\Sigma\chi^2 = 56.51$; $df = 31$; $p < 0.01$). Calculation of standardized residuals for burned flint microartifacts of different excavated units suggested significant values (SR > 2) for only four sub-squares, and in three of those N [expected] < 5. The single sub-square that displays a truly significant SR value is the one encircling the kernel of the southern concentration; there the observed percentage of burned flint microartifacts is 10.00% higher than the expected one (SR = 2.80; N [expected] = 8.72).

Observed percentages of burning exceed the expected ones in additional excavated units. In the eastern part of the exposed surface two sub-squares exhibit a 4.00–4.50% deviation between the observed and expected percentages; however, only five burned flint microartifacts are recorded in each of these sub-squares and the N [expected] < 2 (Fig. 3.9).

Thus the burned flint microartifacts display two adjacent high-density concentrations in the central part of the excavated surface. The kernel of the larger southern concentration covers an area of 0.13 m^2 in which 9.63% of the burned and 4.90% of the unburned flint microartifacts are included; the kernel of the smaller, northern concentration covers 0.022 m^2 and includes 4.81% of the burned and 0.70% of the unburned flint microartifacts.

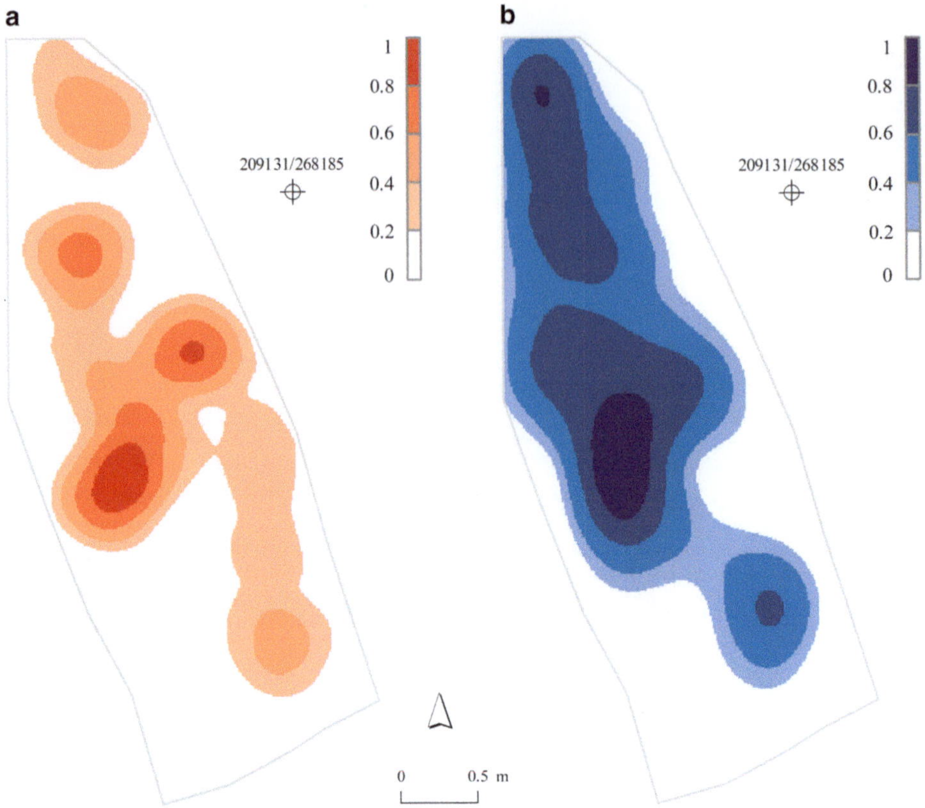

Fig. 3.7 Kernel density maps of flint microartifacts in Layer V-6. (**a**) Burned flint microartifacts (N = 83) and (**b**) unburned flint microartifacts (N = 4,424)

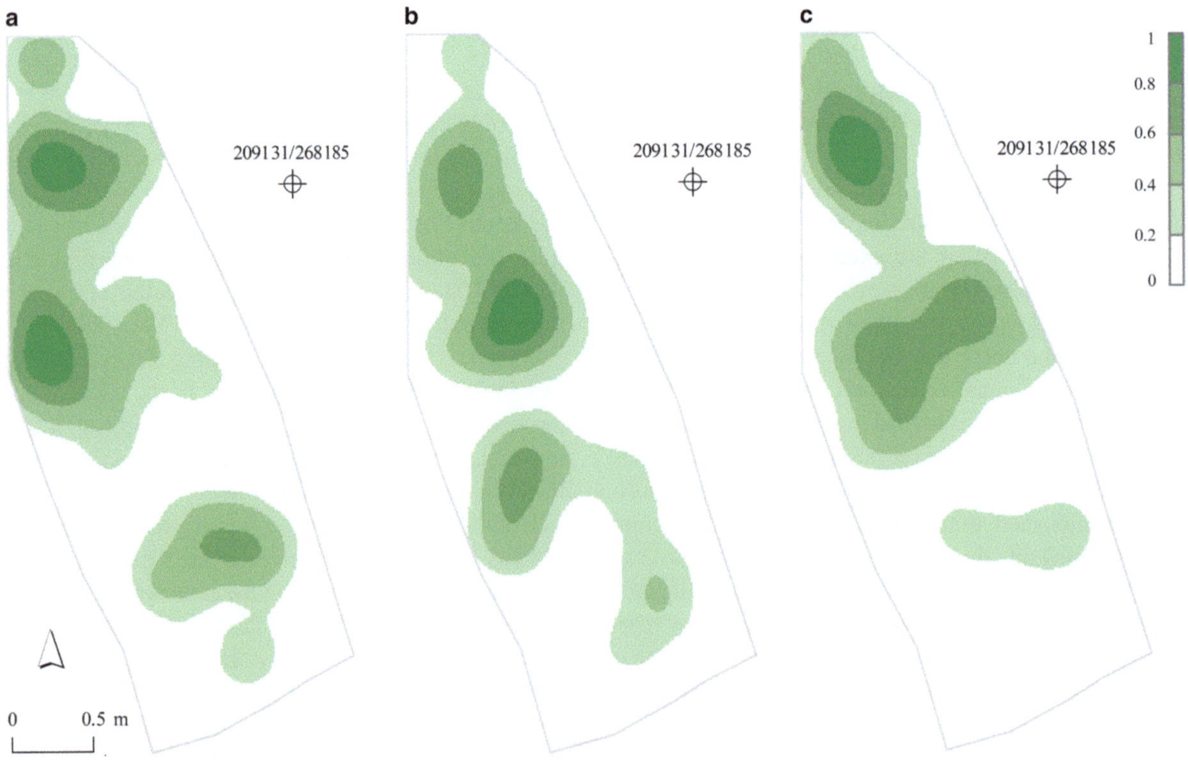

Fig. 3.8 Kernel density maps of three randomly selected data sets (N = 83) for Layer V-6

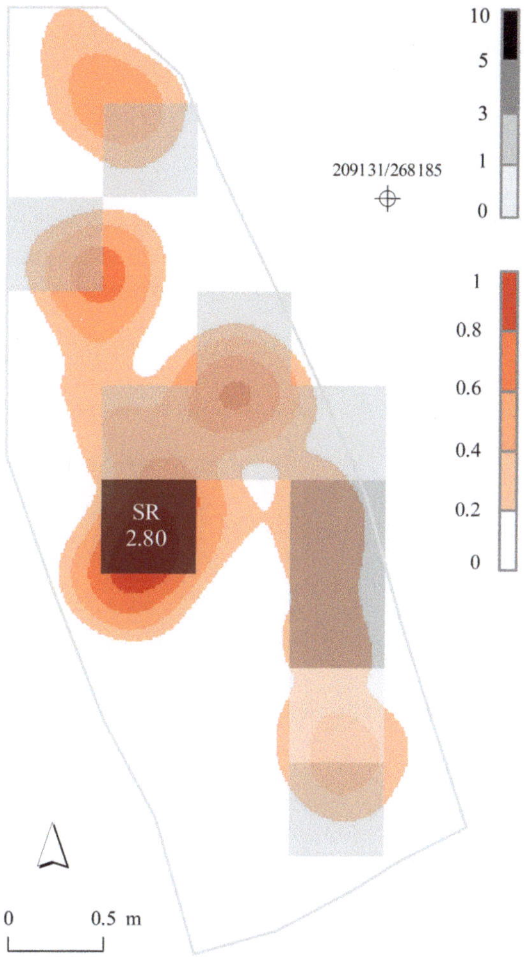

Fig. 3.9 The kernel density map of burned flint microartifacts of Layer V-6 and excavated units in which the observed percentage of burning exceeds the expected percentage; significant SR value is marked on the map

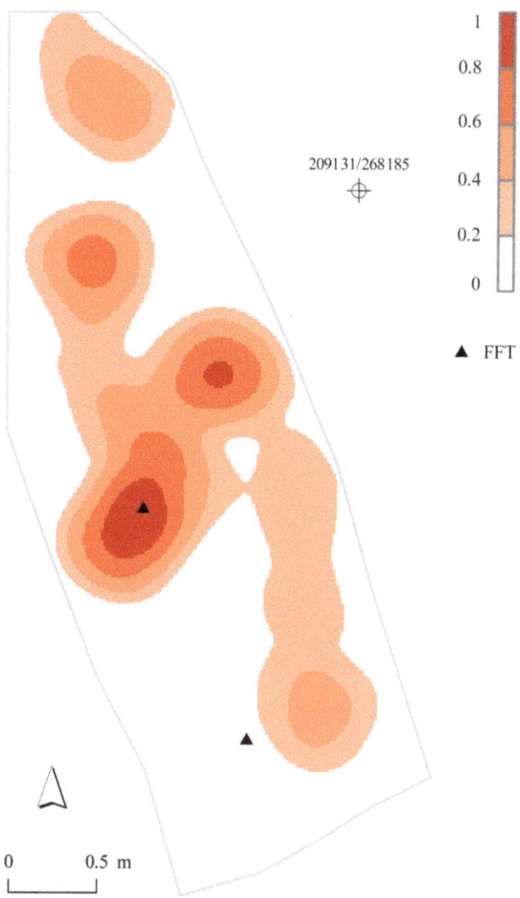

Fig. 3.10 The kernel density map of burned flint microartifacts of Layer V-6 and the distribution of large burned flint items (FFT: N = 2)

Two burned flint macroartifacts were found in Layer V-6 (Table 3.2); one is recorded within the kernel of the southern high-density concentration of burned flint microartifacts, and the second occurs some 2 m to the southeast (Fig. 3.10).

3.3 Layer I-4

For a variety of reasons, specified in the previous chapter (under "Database Construction"), the lithic material from Area A (i.e., Layers I-4 and I-5) is presented only concisely here. The general lithic inventory of Layer I-4 is presented in Table 3.3, which includes the entire lithic assemblage retrieved from that layer.

Spatial data are presented only for the northern part of the excavated area, where the 5.25 m² of exposed surface yielded 6,696 microartifacts and 32 macroartifacts. Within the lithic inventory of this area (Table 3.4), flint is the dominant raw material, exhibiting evidence of burning on 0.61% of the microartifacts (Table 3.4).

Table 3.3 Lithic assemblage of Layer I-4, total assemblage (South and North areas)

Category	Flint				Basalt	Lime-stone	Total
	Unburned		Burned				
	N	%	N	%	N	N	N
Microartifacts[a]	7,402	99.36	47	0.63	106	103	7,658
FFT artifacts	32	100.00	–		4	–	36
Total	7,434	99.37	47	0.62	110	103	7,694

[a] The percentage of burned and unburned flint items is calculated within each lithic category

Table 3.4 Lithic assemblage of Layer I-4 (North), spatially presented

Category	Flint				Basalt	Lime-stone	Total
	Unburned		Burned				
	N	%	N	%	N	N	N
Microartifacts[a]	6,457	99.38	40	0.61	103	96	6,696
FFT artifacts	29	100.00	–		3	–	32
Total	6,486	99.38	40	0.61	106	96	6,728

[a] The percentage of burned and unburned flint items is calculated within each lithic category

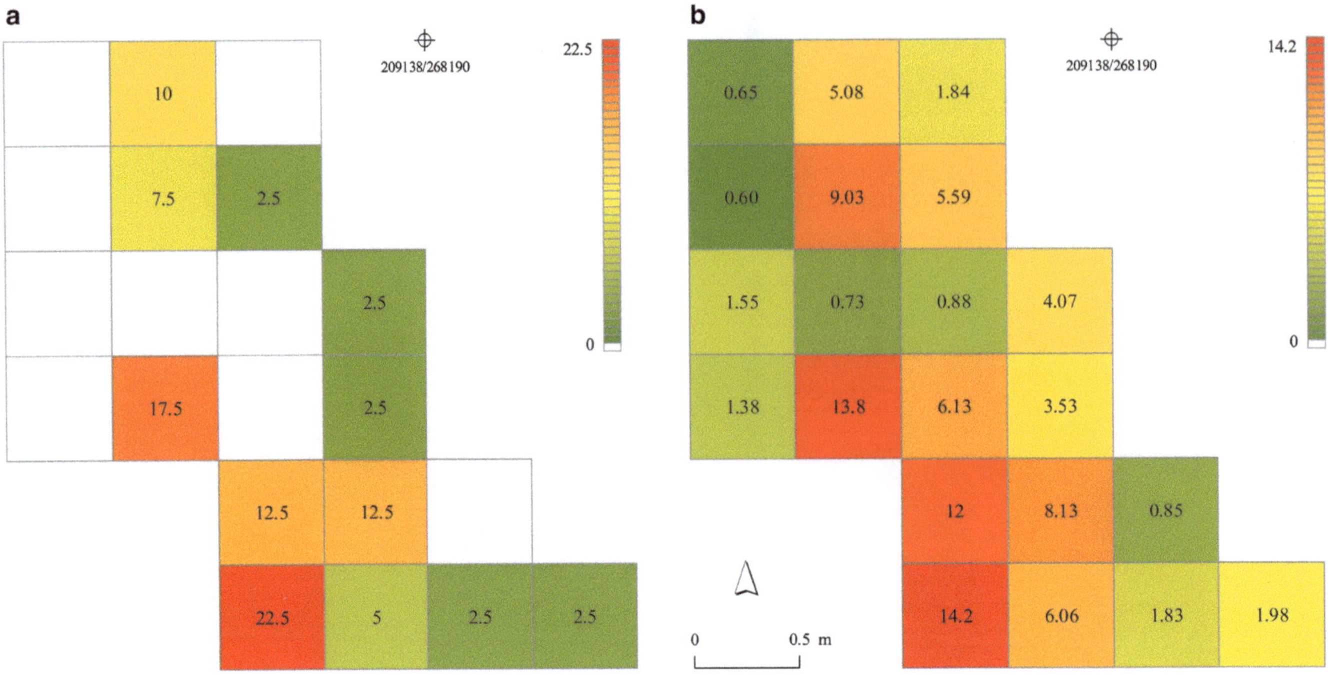

Fig. 3.11 Percentages of flint microartifacts per excavated unit in Layer I-4 (North). (**a**) Burned flint microartifacts (N = 40) and (**b**) unburned flint microartifacts (N = 6,457)

Unburned flint microartifacts were present in each of the excavated units, while the burned ones were retrieved from only 3 of the 5.25 m² of excavated area (Fig. 3.11).

High percentages of burning are observed in the southern half of the exposed area, where over 50% of the burned items occur in an area of 1 m²; slightly to the northwest, an additional 17.5% occurs within a single sub-square. These excavated units, in which the percentage of burned flint microartifacts is relatively high, correspond to the high-percentages units of the unburned flint microartifacts (Fig. 3.11).

Table 3.5 Lithic assemblage of Layer I-5, total assemblage (South and North areas)

Category	Flint				Basalt	Limestone	Total
	Unburned		Burned				
	N	%	N	%	N	N	N
Microartifacts[a]	15,661	99.23	121	0.76	219	135	16,136
FFT artifacts	54	100.00	–		2	–	56
CCT artifacts	9	100.00	–		4	1	14
Total	15,724	99.23	121	0.76	225	136	16,206

[a] The percentage of burned and unburned flint items is calculated within each lithic category

Table 3.6 Lithic assemblage of Layer I-5 (North), spatially presented

Category	Flint				Basalt	Limestone	Total
	Unburned		Burned				
	N	%	N	%	N	N	N
Microartifacts[a]	14,894	99.19	121	0.80	210	125	15,350
FFT artifacts	51	100.00	–		1	–	52
CCT artifacts	9	100.00	–		1	1	11
Total	14,954	99.19	121	0.80	212	126	15,413

[a] The percentage of burned and unburned flint items is calculated within each lithic category

3.4 Layer I-5

As for Layer I-4, the characteristics and spatial distribution of the lithic material from Layer I-5 are briefly discussed here. Table 3.5 incorporates the entire lithic inventory of the layer, while the data in Table 3.6 include only the material from the northern part of the excavated area, for which the general distribution of the flint microartifacts is presented.

The northern area, measuring 5 m², yielded 15,350 microartifacts and 63 macroartifacts of various raw materials, modified predominantly on flint; evidence of burning was observed only on 0.80% of the flint microartifacts (Table 3.6).

The unburned flint microartifacts occurred within each of the excavated units, and the burned flint microartifacts were retrieved from 4.5 of the 5 m² excavated area (Fig. 3.12). High percentages of burning are recorded in several excavated units throughout the exposed area. Similarly, the unburned flint microartifacts do not exhibit clustered areas of high frequencies and are scattered over the entire excavated surface (Fig. 3.12).

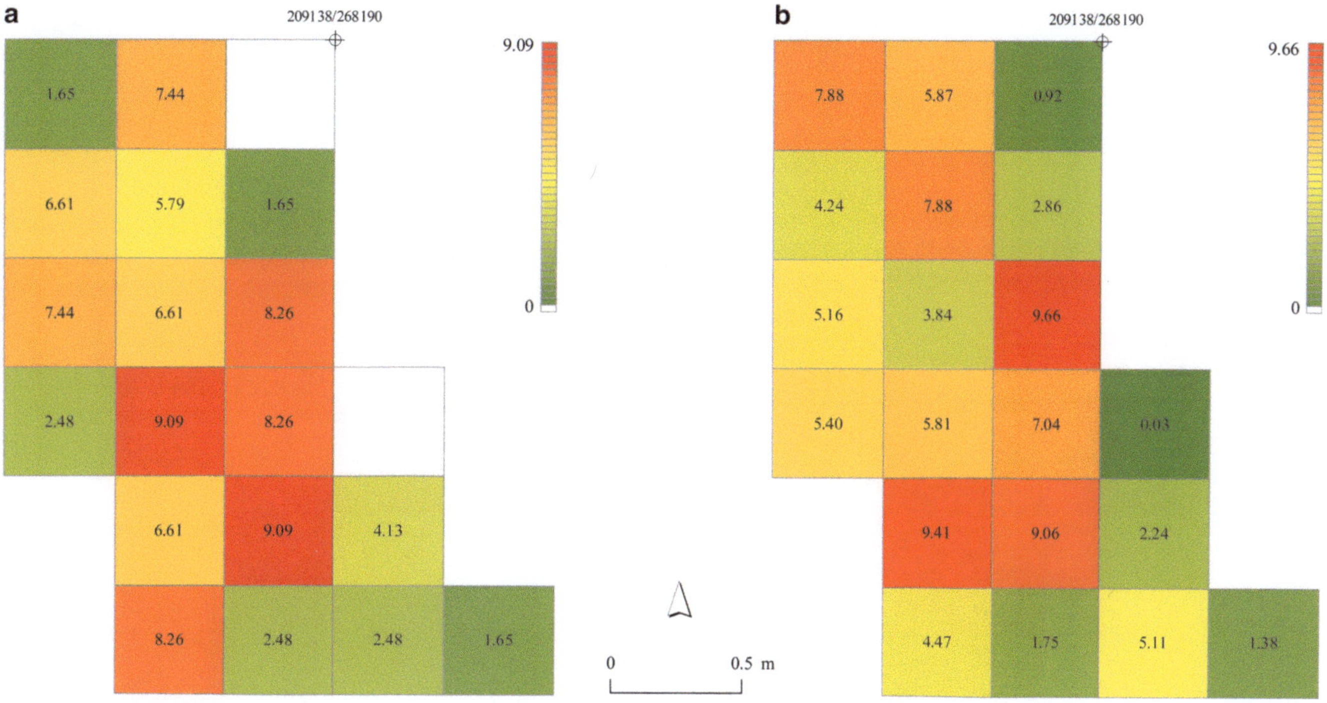

Fig. 3.12 Percentages of flint microartifacts per excavated unit in Layer I-5 (North). (**a**) Burned flint microartifacts (N = 121) and (**b**) unburned flint microartifacts (N = 14,894)

3.5 Layer II-2/3

The data for Layer II-2/3 are presented only briefly and include the general lithic inventory and schematic illustrations of the spatial distribution of the burned and unburned flint. The small spatial extent of this layer does not permit a meaningful spatial analysis. Layer II-2/3 was excavated to an extent of 4.67 m^2, yielding an assemblage of 7,502 microartifacts and 139 macroartifacts of various raw materials (Table 3.7). Burning is seen in very low frequencies on flint microartifacts (0.69%) and macroartifacts (0.94%).

The burned and unburned flint microartifacts appear to be distributed similarly and areas of high frequencies seem to overlap each other. Burned flint microartifacts occur in 10 excavated units and display relatively high frequencies in four sub-squares (Fig. 3.13). High frequencies of unburned flint microartifacts are similarly recorded in these four sub-squares (Fig. 3.13).

A single burned macroartifact was recovered from Layer II-2/3 (Table 3.7). This burned artifact occurs within a sub-square in which relatively high frequencies of burned flint microartifacts are recorded (Fig. 3.13).

Table 3.7 Lithic assemblage of Layer II-2/3

Category	Flint				Basalt	Lime-stone	Total
	Unburned		Burned				
	N	%	N	%	N	N	N
Microartifacts[a]	7,323	99.30	51	0.69	43	85	7,502
FFT artifacts	85	100.00	–		16	2	103
CCT artifacts[a]	20	95.23	1	4.76	9	–	30
Handaxes	–		–		6	–	6
Cleavers	–		–		–	–	–
Pebbles	251	100.00	–		117	51	419
Total	7,679	99.32	52	0.67	191	138	8,060

[a] The percentage of burned and unburned flint items is calculated within each lithic category

3.6 Layer II-5

Layer II-5 was excavated to an extent of 25 m^2, which yielded an assemblage of 3,903 microartifacts and 180 macroartifacts of various raw materials, predominantly flint (Table 3.8). Burning is seen in somewhat higher frequencies on the flint microartifacts (4.48%) than on the flint macroartifacts (3.41%) (Table 3.8).

Excavated units in which burned flint microartifacts occur encompass only 25 of the 117 excavated sub-squares. In addition, their spatial distribution is not uniform. Some 55% of the burned flint microartifacts cover an area of 3.25 m^2 in the northern part of the exposed surface, while the remaining 45% are scattered throughout the excavated area (Fig. 3.14).

The unburned flint microartifacts display similar patterning, with the highest frequencies occurring in the northern part of the excavated surface. However, the unburned flint

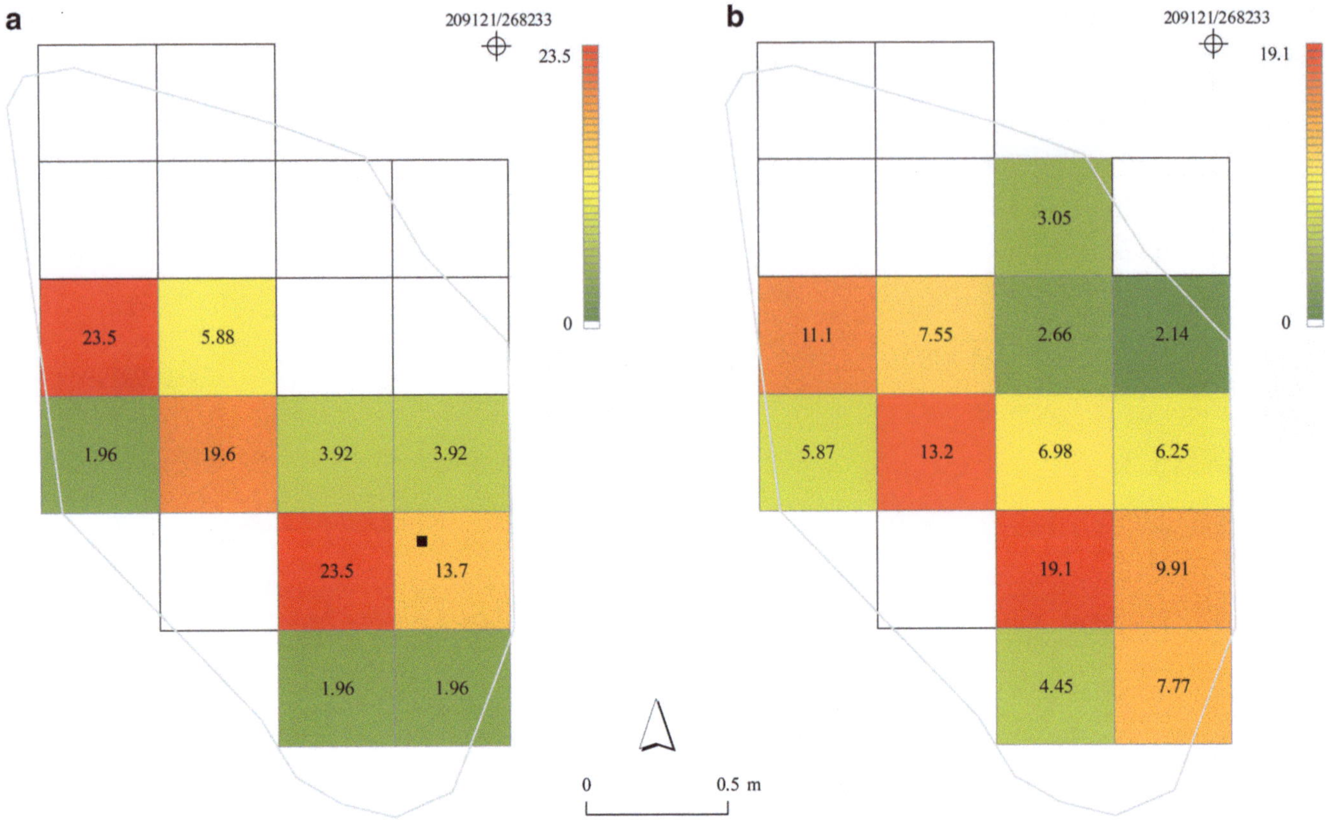

Fig. 3.13 Percentages of flint microartifacts per excavated unit in Layer II-2/3. (**a**) Burned flint microartifacts (N = 51); location of burned flint macroartifact is marked on the map (CCT: N = 1) and (**b**) unburned flint microartifacts (N = 7,323)

Table 3.8 Lithic assemblage of Layer II-5

Category	Flint				Basalt	Lime-stone	Total
	Unburned		Burned				
	N	%	N	%	N	N	N
Microartifacts[a]	3,453	95.51	162	4.48	76	212	3,903
FFT artifacts[a]	92	97.87	2	2.12	47	2	143
CCT artifacts	21	91.30	2	8.69	5	3	31
Handaxes	–		–		4	–	4
Cleavers	–		–		2	–	2
Pebbles	72	100.00	–		135	15	222
Total	3,638	95.63	166	4.36	269	232	4,305

[a] The percentage of burned and unburned flint items is calculated within each lithic category

microartifacts cover a larger surface and are present in most of the excavated units (Fig. 3.14).

The dense concentration in the northern part of the excavated area is further illustrated in the density maps, which suggest comparable patterns of distribution for the burned and unburned flint microartifacts (Fig. 3.15).

The random density maps follow this general overlapping and point to the northern area as the area most likely to display high density of burning (Fig. 3.16).

The overlapping between the burned and unburned flint microartifacts is, however, not complete. As mentioned above, the burned flint microartifacts cover a significantly smaller area. In addition, within the northern area of 3.25 m^2 in which the burned flint microartifacts are abundant, several excavated units display higher percentages of burning than what we would expect if the distribution were uniform (Fig. 3.17). The highest deviation between the observed and expected percentages of burning (2.96%) is recorded in a sub-square within the northern concentration (SR = 1.50; N [expected] = 10.2) (Fig. 3.17).

The burned flint microartifacts are clustered and display a dense concentration (with over 50% of the burned items) in the northern part of the excavated surface. Correspondingly, applying a chi square test for the burned flint microartifacts suggests that the distribution of these is significantly different from an expected, uniform distribution ($\Sigma\chi^2 = 200.43$; $df = 73$; $p < 0.001$). Calculation of standardized residuals for burned flint microartifacts of different excavated units suggests several sub-squares as potential contributors to the observed clustering (i.e., SR > 2); however, none of these sub-squares displays sufficient numbers (i.e., N [expected] > 5) of burned flint microartifacts.

Within the kernel of the cluster of burned flint microartifacts, which covers an area of 0.219 m^2, the relative percentage of the burned flint microartifacts (10.49%) is higher than that of the unburned ones (7.73%).

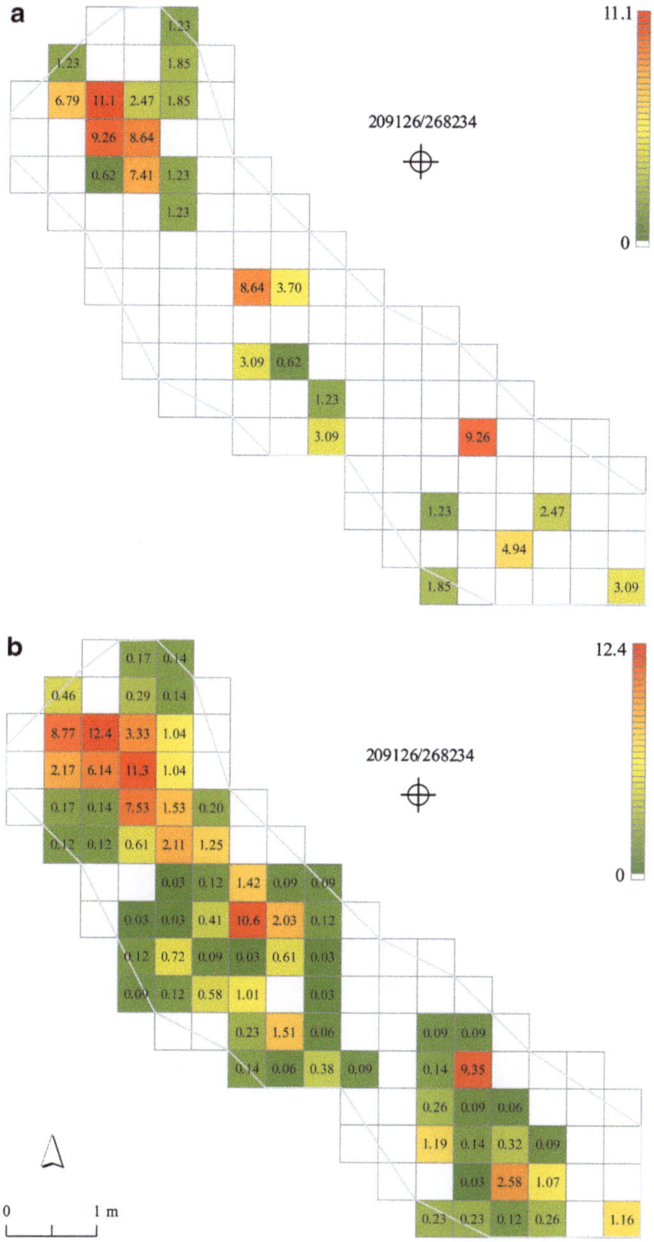

Fig. 3.14 Percentages of flint microartifacts per excavated unit in Layer II-5. (**a**) Burned flint microartifacts (N = 162) and (**b**) unburned flint microartifacts (N = 3,453)

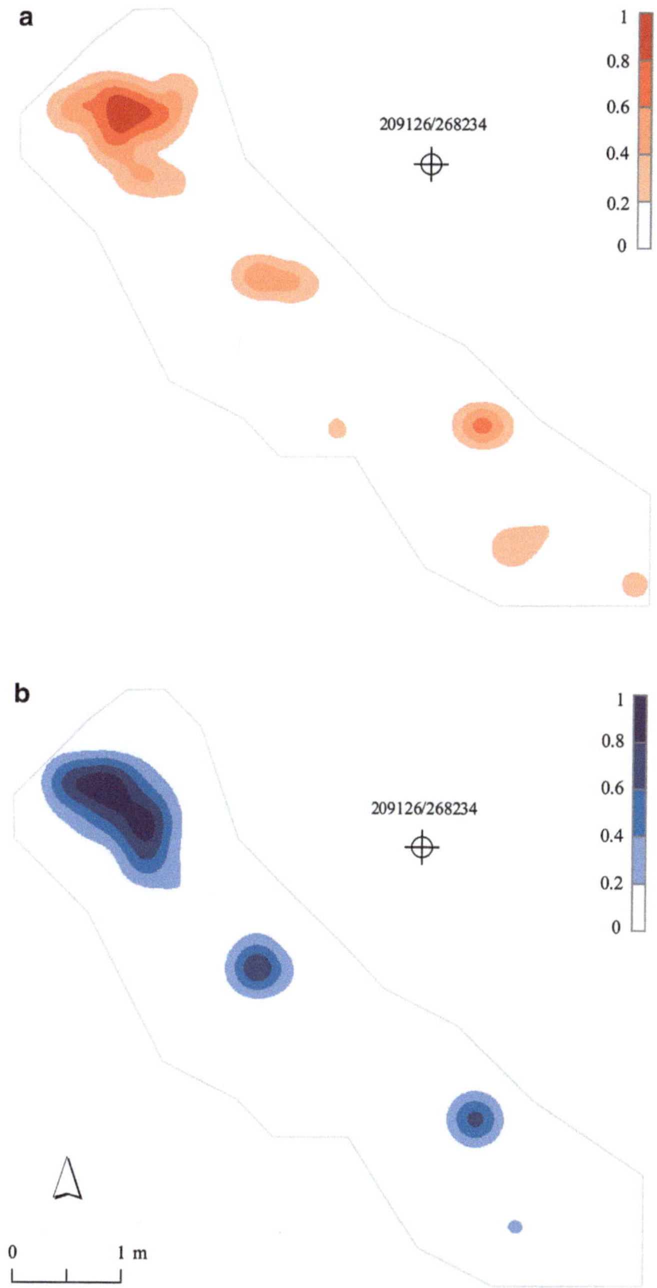

Fig. 3.15 Kernel density maps of flint microartifacts in Layer II-5. (**a**) Burned flint microartifacts (N = 162) and (**b**) unburned flint microartifacts (N = 3,453)

The flint assemblage of Layer II-5 yielded four burned macroartifacts (Table 3.8). Three of these occur within and in close vicinity to the high-density concentration of burned flint microartifacts; the fourth is located some 6 m to the southeast of the concentration (Fig. 3.18).

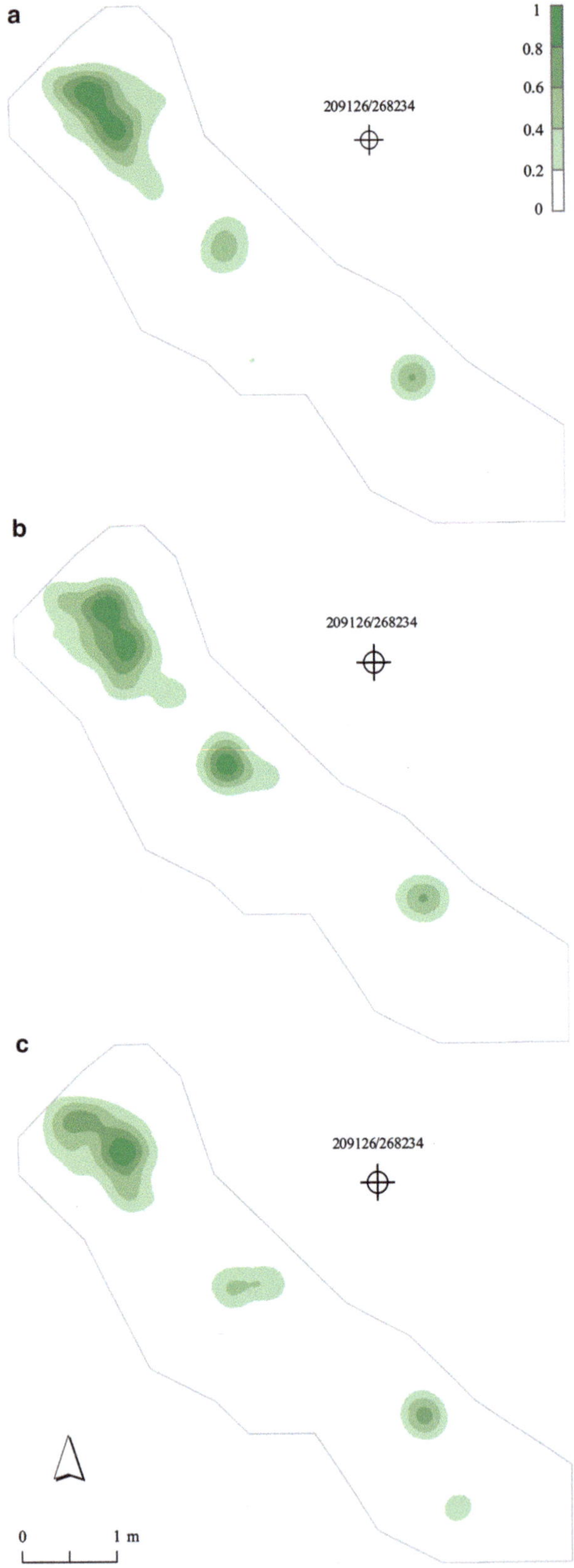

Fig. 3.16 Kernel density maps of three randomly selected data sets (N = 162) for Layer II-5

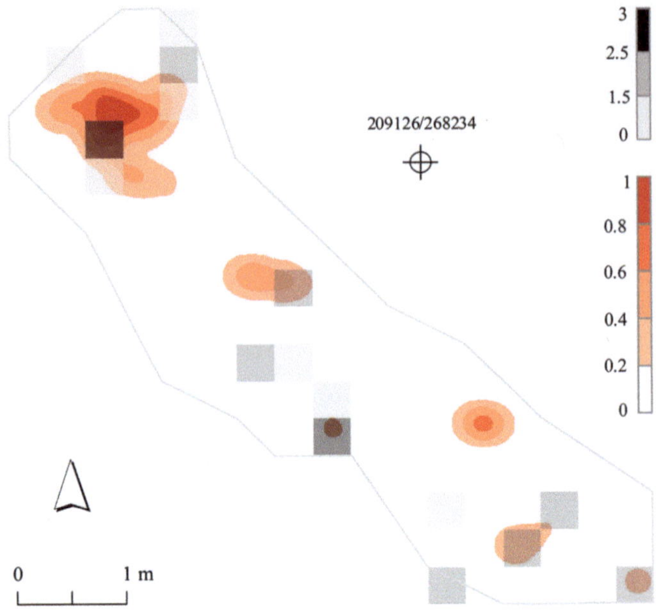

Fig. 3.17 The kernel density map of burned flint microartifacts of Layer II-5 and excavated units in which the observed percentage of burning exceeds the expected percentage

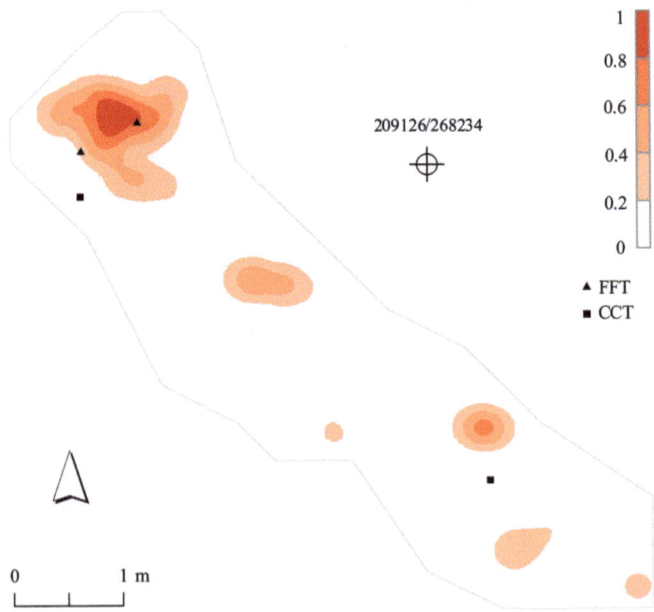

Fig. 3.18 The kernel density map of burned flint microartifacts of Layer II-5 and the distribution of large burned flint items (FFT: N = 2; CCT: N = 2)

3.7 Layer II-5/6

The lithic assemblage of Layer II-5/6 consists of 10,531 microartifacts and 142 macroartifacts, recovered from an area of 19.14 m². Flint is the dominant raw material, exhibiting relatively high frequencies of burning amongst microartifacts (4.50%) and macroartifacts (5.30%) (Table 3.9).

Burned flint microartifacts occur in 37 of the 93 excavated sub-squares. Two adjacent areas, in the central part of the excavated surface, display high frequencies of burned flint microartifacts. The first (southern) covers 1.75 m² and includes 25.75% of the burned flint microartifacts; the second (slightly to the northwest) incorporates 45.25% of the burned flint microartifacts of the layer within an area of 3.5 m² (Fig. 3.19).

The distribution of the unburned flint microartifacts covers a larger area, with most of the excavated units being represented. Similarly to the burned flint microartifacts, the highest frequencies are observed in two adjacent areas: in the center of the excavated surface (some 20% of the unburned flint microartifacts) and slightly to the northwest (some 52% of the unburned flint microartifacts) (Fig. 3.19).

These patterns are further illustrated in the density maps of the flint microartifacts (Fig. 3.20). The density map of the unburned flint microartifacts emphasizes the high-density concentration in the northwestern area, while the central concentration exhibits lower levels of density (Fig. 3.20). The central concentration appears to be more definite in the density map of the burned flint microartifacts, which display high densities in the northwestern as well as the central areas (Fig. 3.20). Following these patterns, the random density maps give more emphasis to the northwestern area as the area most likely to exhibit high frequencies of burning (Fig. 3.21).

Layer II-5/6 thus displays two concentrations of burned flint microartifacts, in the northwest (henceforth: northern) and in the center (henceforth: southern) of the excavated surface. The chi square test applied to the burned flint microartifacts suggests that the distribution of these is significantly different from an expected, uniform distribution ($\Sigma\chi^2 = 204.48$; $df = 64$; $p < 0.001$). Calculation of standardized residuals for

Table 3.9 Lithic assemblage of Layer II-5/6

Category	Flint				Basalt	Lime-stone	Total
	Unburned		Burned				
	N	%	N	%	N	N	N
Microartifacts[a]	8,487	95.49	400	4.50	589	1,055	10,531
FFT artifacts[a]	95	94.05	6	5.90	24	–	125
CCT artifacts	12	100.00	–		1	2	15
Handaxes	–		–		1	–	1
Cleavers	–		–		1	–	1
Pebbles	108	100.00	–		139	19	266
Total	8,702	95.54	406	4.45	755	1,076	10,939

[a] The percentage of burned and unburned flint items is calculated within each lithic category

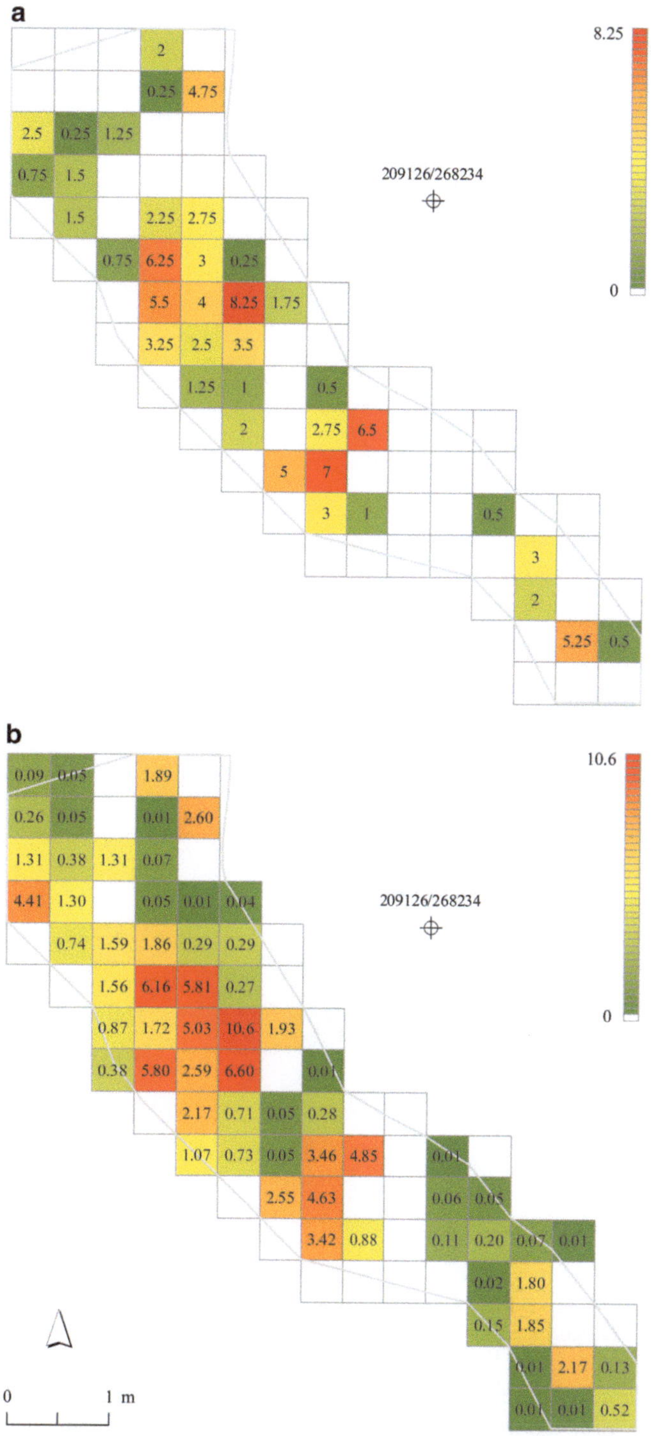

Fig. 3.19 Percentages of flint microartifacts per excavated unit in Layer II-5/6. (**a**) Burned flint microartifacts (N = 400) and (**b**) unburned flint microartifacts (N = 8,487)

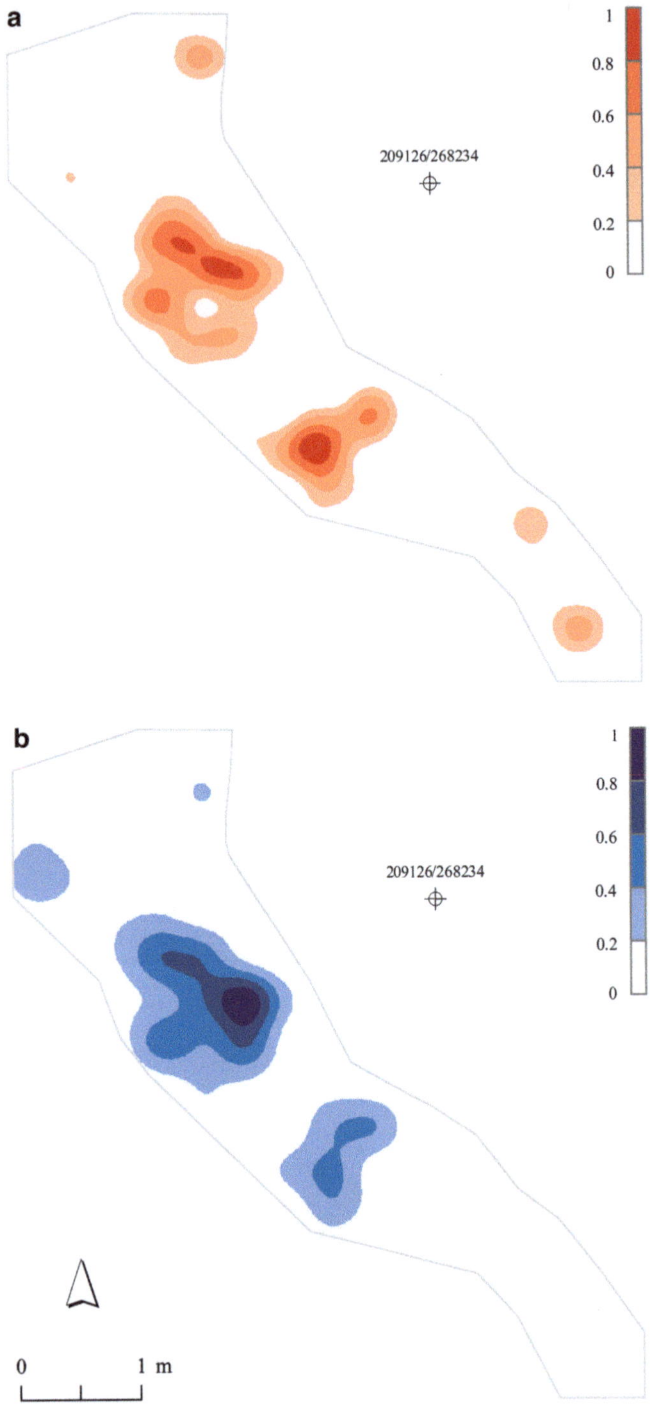

Fig. 3.20 Kernel density maps of flint microartifacts in Layer II-5/6. (**a**) Burned flint microartifacts (N = 400) and (**b**) unburned flint microartifacts (N = 8,487)

burned flint microartifacts of different excavated units suggests that significant values (SR > 2), pointing to the potential contributors to the observed clustering, are observed in sub-squares within and in the vicinity of the burned clusters (specified below).

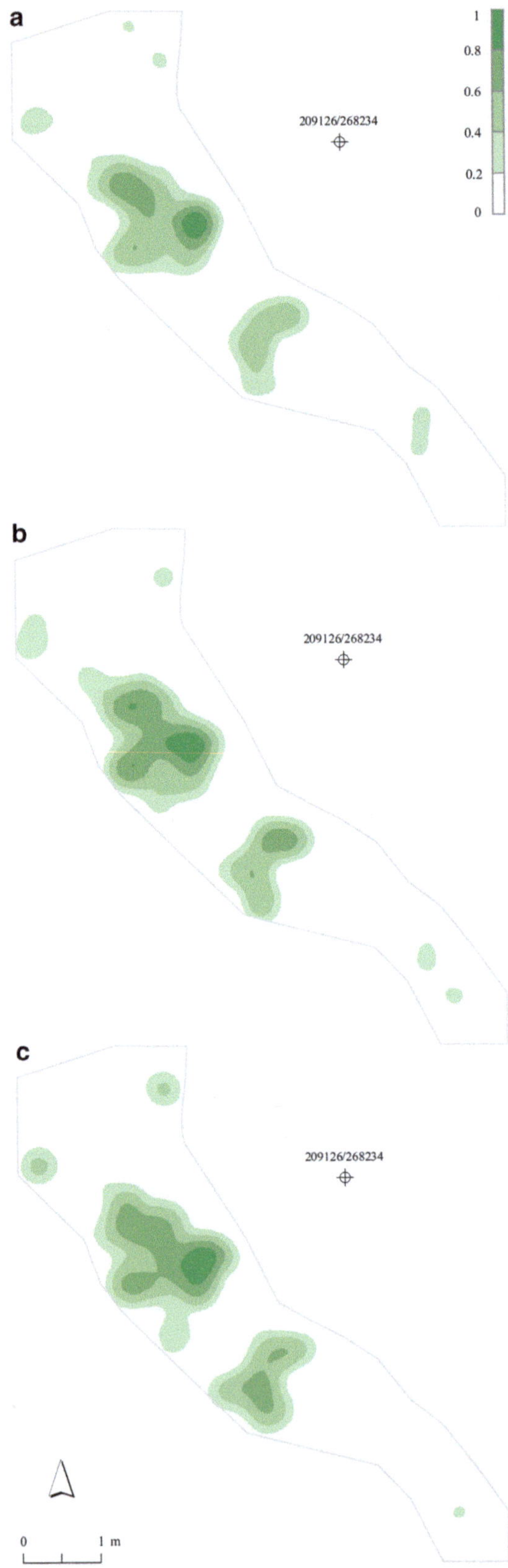

Fig. 3.21 Kernel density maps of three randomly selected data sets (N = 400) for Layer II-5/6

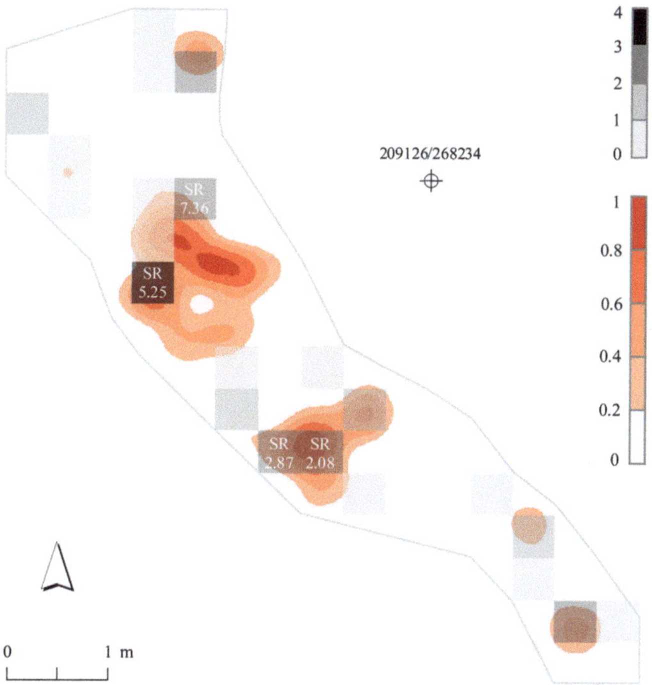

Fig. 3.22 The kernel density map of burned flint microartifacts of Layer II-5/6 and excavated units in which the observed percentage of burning exceeds the expected percentage; significant SR values are marked on the map

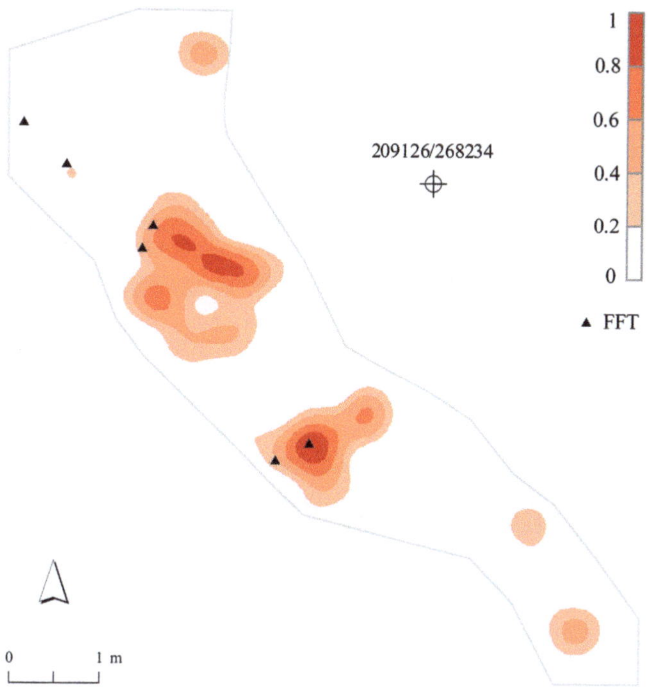

Fig. 3.23 The kernel density map of burned flint microartifacts of Layer II-5/6 and the distribution of large burned flint items (FFT: N = 6)

In the northern concentration, several excavated units display higher percentages of burning than what we would expect if the distribution of the burned flint microartifacts were uniform (Fig. 3.22). Particularly, in the sub-square that displays the highest deviation between the observed and expected values of burning (3.60%) the SR value is significant (SR = 5.25; N [expected] = 7.56). The sub-square slightly to its northeast, where the observed percentage of burned flint microartifacts is 2.34% higher than the expected one, the SR value is also significant (SR = 7.36; however N [expected] = 1.62).

High density values are represented in the northern concentration in two kernels that together encompass an area of 0.167 m^2 and include 8.00% of the burned and 4.59% of the unburned flint microartifacts of the layer.

In the southern concentration, which is less distinct in the random density maps (Fig. 3.21), several excavated units display higher percentages of burning than what we would expect if the distribution of the burned flint microartifacts were uniform. Significant SR values are observed in two of these sub-squares, which display a gap of 2–3% between the observed and expected percentages of burned flint microartifacts (Fig. 3.22). In the sub-square that encircles the kernel of the concentration, the observed percentage of burning is 2.26% higher than the expected one (SR = 2.08; N [expected] = 18.94) and in the sub-square to its west the observed percentage of burning is 2.34% higher than the expected one (SR = 2.87; N [expected] = 10.62). The kernel of the southern concentration covers an area of 0.122 m^2 and includes 4.50% of the burned and only 1.99% of the unburned flint microartifacts of the layer.

Six burned flint macroartifacts were found in Layer II-5/6 (Table 3.9). Four of these occur within the northern (N = 2) and the southern (N = 2) concentrations of burned flint microartifacts (Fig. 3.23). The remaining two are in the northwestern part of the excavated surface (Fig. 3.23).

3.8 Layer II-6 L-1

Layer II-6 L-1 was exposed over an area of 23.79 m^2, from which an extensive lithic assemblage was recovered (Table 3.10). Flint is the dominant raw material, particularly amongst microartifacts, while within the macroartifacts the dominance of flint is closely followed by basalt, exhibiting quite similar frequencies (Table 3.10).

The percentage of burned flints is relatively low amongst microartifacts (1.40%) and macroartifacts (1.40%). A particularly low percentage of burning (0.25%) is observed for flint pebbles (Table 3.10).

Table 3.10 Lithic assemblage of Layer II-6 L-1

Category	Flint				Basalt	Lime-stone	Total
	Unburned		Burned				
	N	%	N	%	N	N	N
Microartifacts[a]	53,081	98.59	754	1.40	2,745	1,506	58,086
FFT artifacts[a]	781	99.11	7	0.88	869	22	1,679
CCT artifacts[a]	407	97.60	10	2.39	114	16	547
Handaxes	4	100.00	–		46	4	54
Cleavers	–		–		15	–	15
Pebbles[a]	2,339	99.74	6	0.25	3,867	366	6,578
Total	56,612	98.64	777	1.35	7,656	1,914	66,959

[a] The percentage of burned and unburned flint items is calculated within each lithic category

Burned flint microartifacts are distributed throughout the excavated area and are represented in 69 of the 109 excavated sub-squares. Two areas, in the central and northwestern parts of the exposed surface, exhibit higher frequencies of burned microartifacts (Fig. 3.24). The spatial distribution of the unburned flint microartifacts, however, is different. These are scattered throughout most of the excavated surface, displaying areas of high frequencies in several excavated units extending from the southeastern corner of the exposed surface to the northwestern one (Fig. 3.24).

The density map of unburned flint microartifacts illustrates that the highest density values occur in the southeastern corner of the excavated surface (Fig. 3.25). The density map of the burned flint microartifacts illustrates two concentrations of high density; one in the northwestern corner of the excavated unit and another slightly south of it (Fig. 3.25).

The various distribution and density maps suggest that the spatial patterning of the burned and unburned flint microartifacts of Layer II-6 L-1 do not coincide. Rather, they display generally opposite patterns, with denser concentrations of burned flint microartifacts in the northwestern part of the excavated surface and of unburned flint microartifacts in the southeast (Fig. 3.25). Correspondingly, random density maps suggest the southeastern corner of the excavated surface as most likely to display high densities of burning (Fig. 3.26).

Layer II-6 L-1 thus displays two concentrations of burned flint microartifacts, in the northwestern corner of the excavated surface (henceforth: northern) and slightly to its south (henceforth: southern). The chi square test of the burned flint microartifacts points to a significant difference between the expected distribution of burning and the observed patterning ($\Sigma\chi^2 = 580.57$; $df = 90$; $p < 0.001$).

In the northern concentration of burned flint microartifacts, several excavated units display higher percentages of burning than what we would expect if the distribution of the burned flint microartifacts were uniform (Fig. 3.27). In the sub-square that encircles the kernel of the concentration, the deviation between the observed and expected values of burning is 2.83% and the SR value is significant (SR = 5.25; N [expected] = 16.6). The kernel of the northern concentration covers an area of 0.148 m^2 and includes 3.97% of the burned flint microartifacts of the layer and 1.25% of the unburned ones.

In the southern concentration of burned flint microartifacts, the sub-square that encircles the kernel exhibits a 3.32% deviation between the observed and expected values of burning and displays a significant SR value (SR = 7.60;

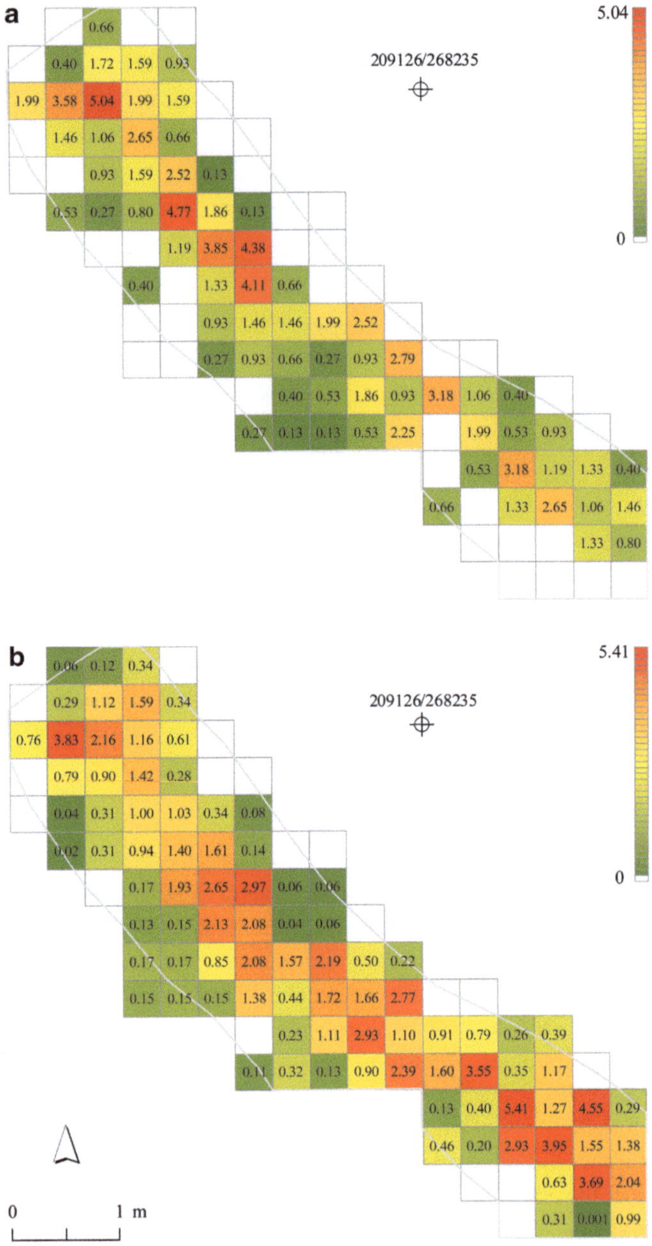

Fig. 3.24 Percentages of flint microartifacts per excavated unit in Layer II-6 L-1. (**a**) Burned flint microartifacts (N = 754) and (**b**) unburned flint microartifacts (N = 53,081)

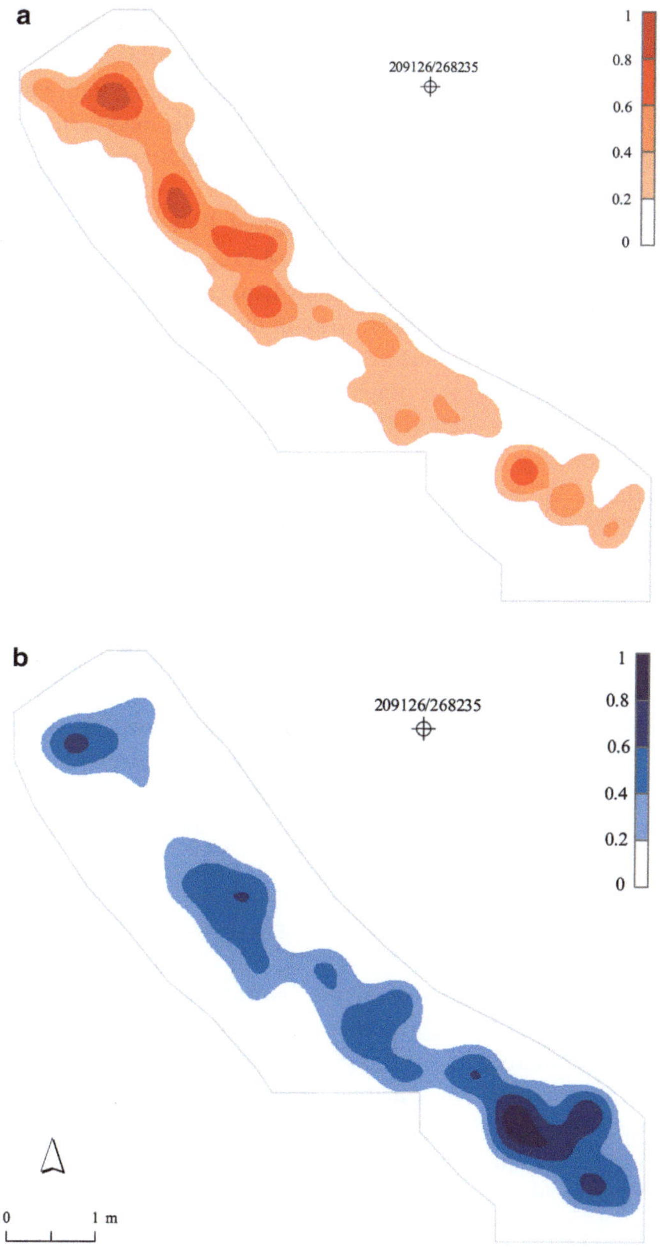

Fig. 3.25 Kernel density maps of flint microartifacts in Layer II-6 L-1. (**a**) Burned flint microartifacts (N = 754) and (**b**) unburned flint microartifacts (N = 53,081)

N [expected] = 10.9). The kernel of the southern concentration covers an area of 0.117 m^2 in which 3.71% of the burned flint microartifacts of the layer and only 0.59% of the unburned ones are included.

Additional burned flint items are found in the lithic assemblage of Layer II-6 L-1 and include six pebbles and 17 macroartifacts (Table 3.10). These seem to be scattered throughout the excavated surface, with only three macroartifacts in close vicinity to the northern concentration of burned flint microartifacts (Fig. 3.28).

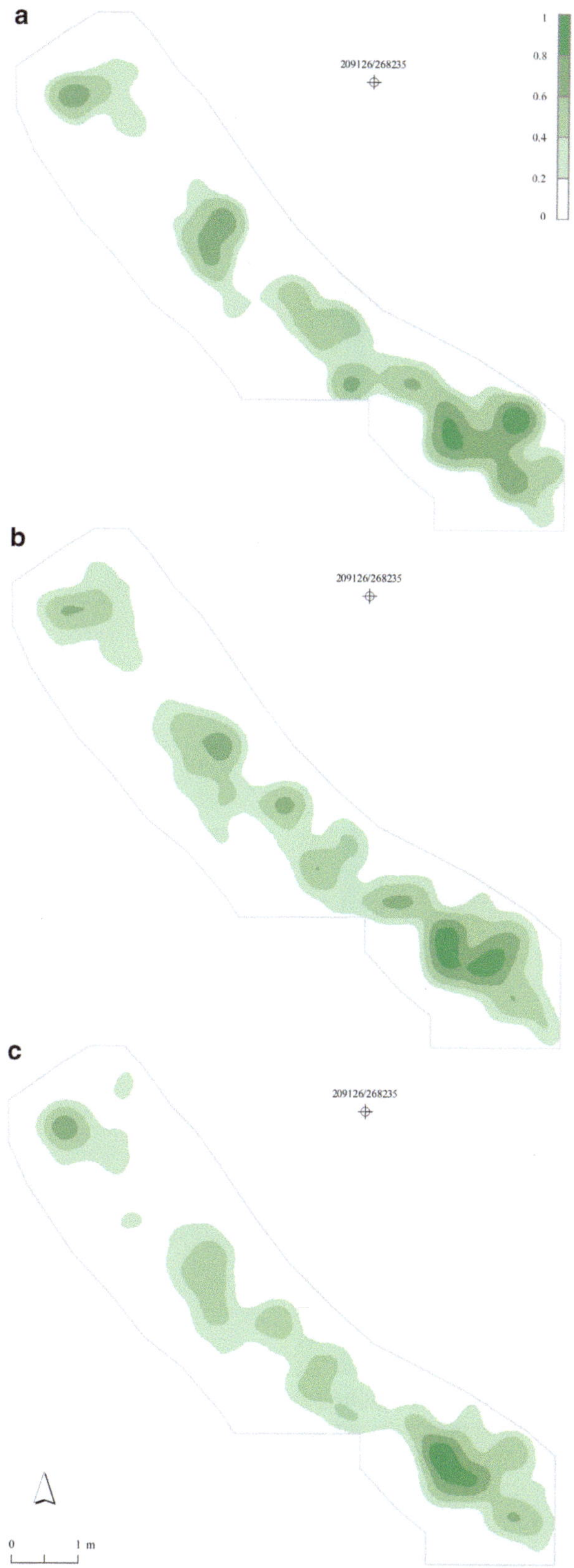

Fig. 3.26 Kernel density maps of three randomly selected data sets (N = 754) for Layer II-6 L-1

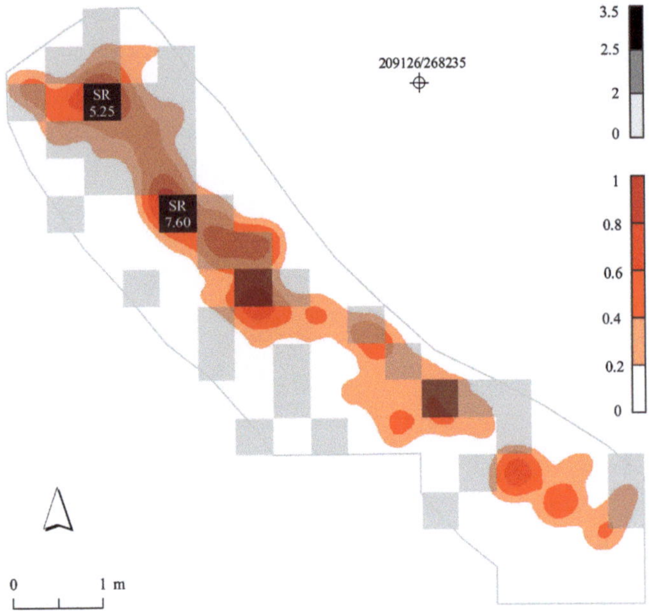

Fig. 3.27 The kernel density map of burned flint microartifacts of Layer II-6 L-1 and excavated units in which the observed percentage of burning exceeds the expected percentage; significant SR values are marked on the map

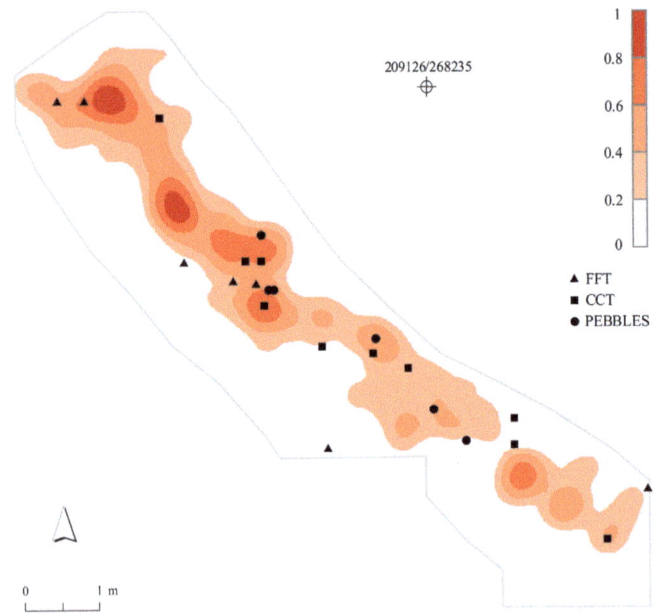

Fig. 3.28 The kernel density map of burned flint microartifacts of Layer II-6 L-1 and the distribution of large burned flint items (FFT: N = 7; CCT: N = 10; Pebbles: N = 6)

3.9 Layer II-6 L-2

A large lithic assemblage, consisting of some 79,670 microartifacts and 1,412 macroartifacts, was recovered from the 25.62 m² exposure of Layer II-6 L-2. Flint is the principal raw material among the microartifacts, while the majority of macroartifacts are modified on basalt (Table 3.11).

Flint exhibits very low frequencies of burning, with only 0.76% of the microartifacts and 1.05% of the macroartifacts being burned. In addition, 0.25% of the flint pebbles show signs of burning (Table 3.11).

Excavated units in which burned flint microartifacts occur consist of only 48 of the 114 excavated sub-squares. Close to 60% of the burned flint microartifacts are concentrated in a 3.25 m² area in the southeastern corner of the excavated surface (Fig. 3.29). Only 22% of the unburned flint microartifacts occur in this 3.25 m² area; they are scattered throughout most of the excavated surface (Fig. 3.29). In contrast with the burned flint microartifacts, the unburned ones exhibit the highest frequencies (57%) in the northwestern corner of the excavated surface (Fig. 3.29).

This divergent patterning is well illustrated in the density maps of the burned and unburned flint microartifacts (Fig. 3.30). The density map of the burned flint microartifacts displays a single high-density concentration in the southeastern corner of the excavated surface. A single high-density concentration is also presented in the density map of the unburned flint microartifacts; however, it is located in the northwestern corner of the excavated area (Fig. 3.30). This spatial deviation is also manifested in the random density maps, which suggest the northwestern part of the excavated surface as most likely to exhibit high density of burning (Fig. 3.31).

The chi square test of the burned flint microartifacts substantiates the significance of the apparent clustering of burned microartifacts ($\Sigma\chi^2 = 913.27$; $df = 68$; $p < 0.001$).

Furthermore, within the concentration of burned flint microartifacts, the percentage of burned flint microartifacts in the sub-square that encircles the highest-density kernel is 12.63% higher than what we would expect if the distribu-

Table 3.11 Lithic assemblage of Layer II-6 L-2

Category	Flint				Basalt	Limestone	Total
	Unburned		Burned				
	N	%	N	%	N	N	N
Microartifacts[a]	73,064	99.23	563	0.76	3,889	2,154	79,670
FFT artifacts[a]	300	99.00	3	0.99	771	15	1,089
CCT artifacts[a]	165	98.80	2	1.19	116	8	291
Handaxes	4	100.00	–		18	–	22
Cleavers	–		–		10	–	10
Pebbles[a]	792	99.74	2	0.25	875	107	1,776
Total	74,325	99.23	570	0.76	5,679	2,284	82,858

[a] The percentage of burned and unburned flint items is calculated within each lithic category

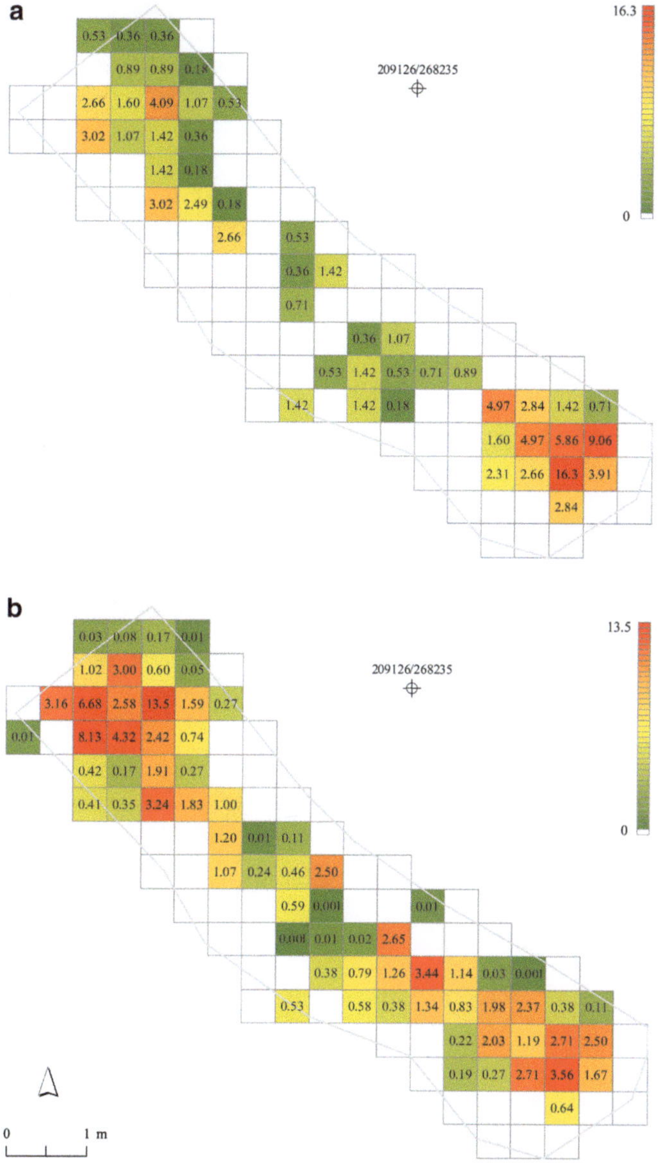

Fig. 3.29 Percentages of flint microartifacts per excavated unit in Layer II-6 L-2. (**a**) Burned flint microartifacts (N = 563) and (**b**) unburned flint microartifacts (N = 73,064)

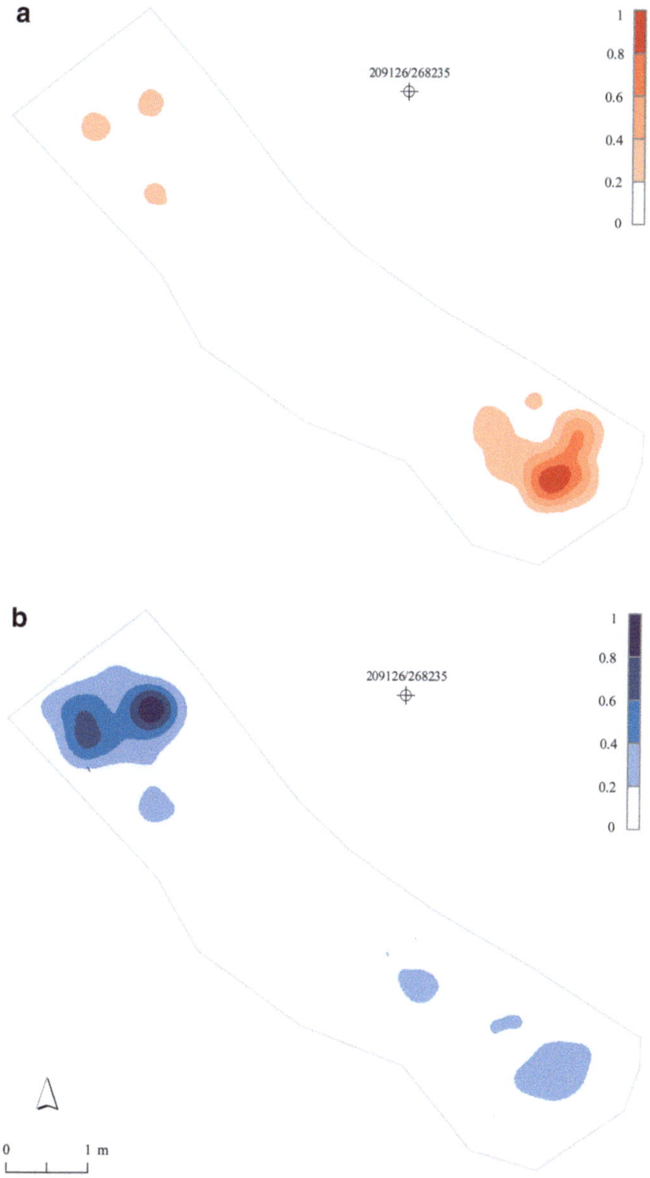

Fig. 3.30 Kernel density maps of flint microartifacts in Layer II-6 L-2. (**a**) Burned flint microartifacts (N = 563) and (**b**) unburned flint microartifacts (N = 73,064)

tion of the burning were uniform (Fig. 3.32). This pattern and the high significance of the standardized residual test of this sub-square (SR = 15.79; N [expected] = 20.5) points to this concentration as a major contributor to the observed clustering. The highest SR values occur in the sub-squares that encircle the concentration of burned flint microartifacts (within the 13 sub-squares included in the cluster, the average SR value is 5.07). The kernel of the southeastern concentration covers an area of 0.165 m^2 and contains 9.41% of the burned flint microartifacts of the layer and 2.39% of the unburned ones.

Five burned flint macroartifacts and two burned flint pebbles are recorded within the lithic assemblage of Layer II-6 L-2 (Table 3.11). Two burned flint macroartifacts occur within the high-density concentration of burned flint microartifacts. The remaining burned flint macroartifacts and the two burned flint pebbles occur mostly in the central part in the excavated surface (Fig. 3.33).

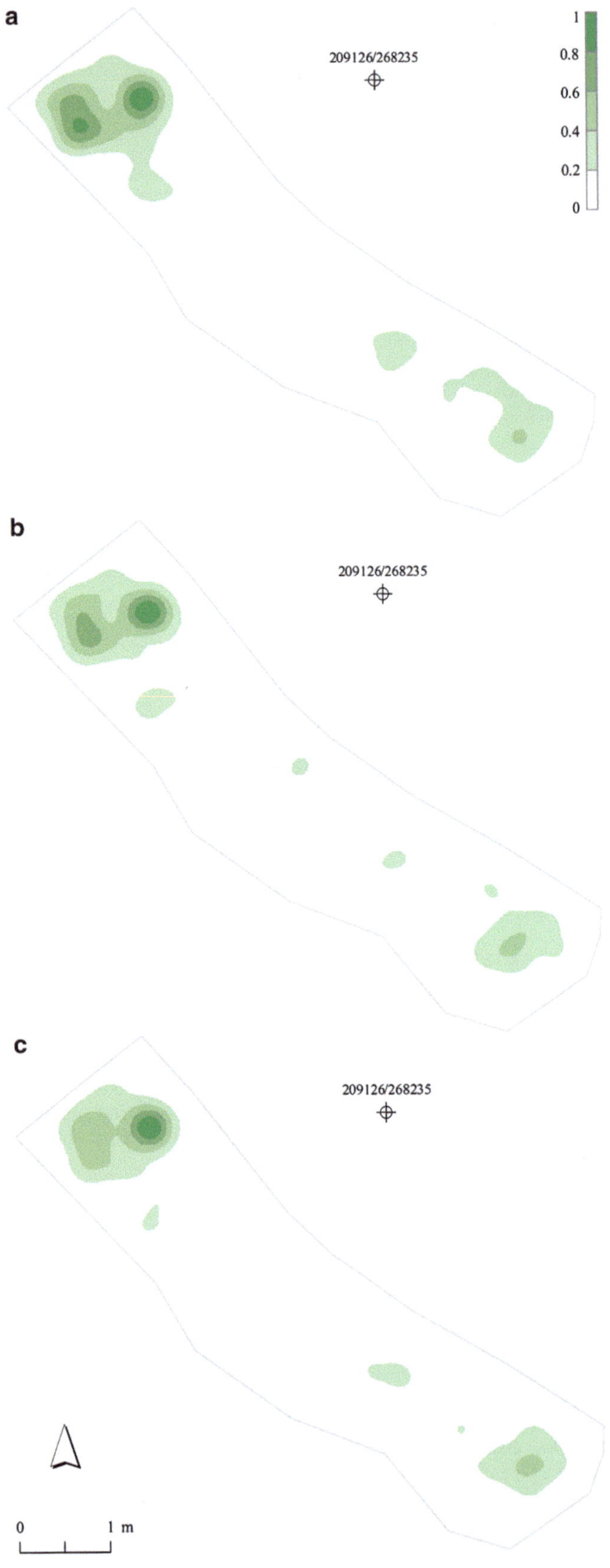

Fig. 3.31 Kernel density maps of three randomly selected data sets (N = 563) for Layer II-6 L-2

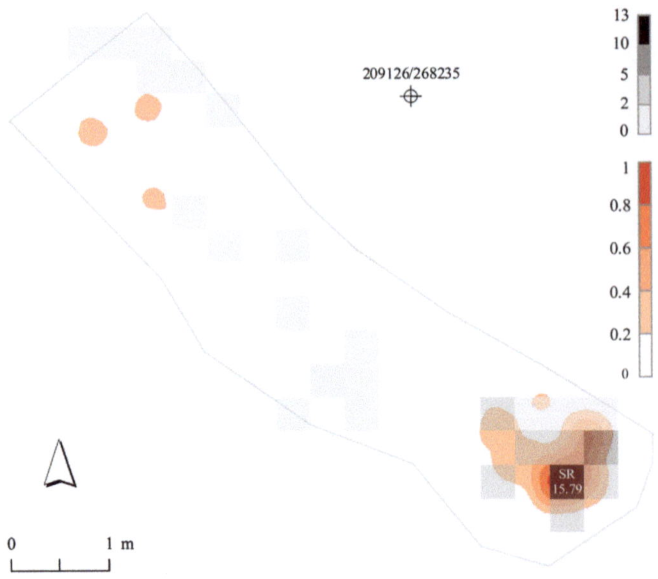

Fig. 3.32 The kernel density map of burned flint microartifacts of Layer II-6 L-2 and excavated units in which the observed percentage of burning exceeds the expected percentage; significant SR value is marked on the map

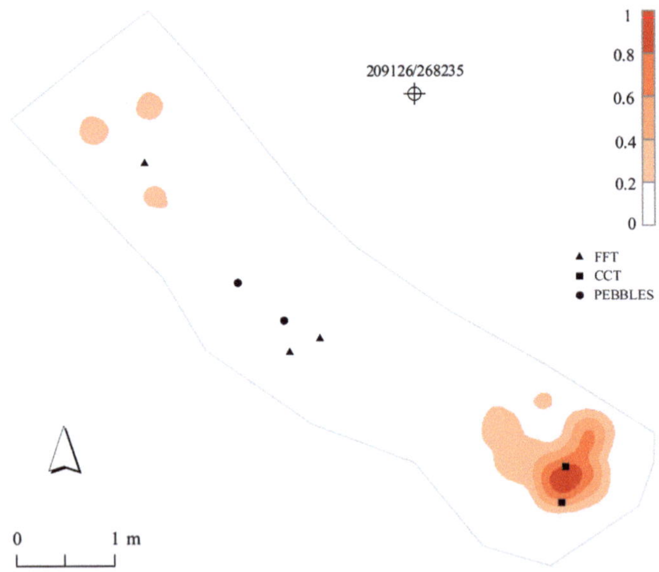

Fig. 3.33 The kernel density map of burned flint microartifacts of Layer II-6 L-2 and the distribution of large burned flint items (FFT: N = 3; CCT: N = 2; Pebbles: N = 2)

3.10 Layer II-6 L-3

Some 17.92 m² of exposed surface of Layer II-6 L3 yielded the large number of 96,094 microartifacts and 1,199 macroartifacts. The microartifacts are predominantly flint, while basalt is more dominant amongst the macroartifacts (Table 3.12). While the flint microartifacts exhibit a low percentage of burning (0.95%), amongst the flint macroartifacts a higher

Table 3.12 Lithic assemblage of Layer II-6 L-3

Category	Flint				Basalt	Lime-stone	Total
	Unburned		Burned				
	N	%	N	%	N	N	N
Microartifacts[a]	91,050	99.04	877	0.95	2,597	1,570	96,094
FFT artifacts[a]	319	98.45	5	1.54	580	21	925
CCT artifacts[a]	175	94.08	11	5.91	58	12	256
Handaxes	–		–		6	–	6
Cleavers	–		–		12	–	12
Pebbles[a]	1,101	99.90	1	0.09	730	154	1,986
Total	92,645	99.04	894	0.95	3,983	1,757	99,279

[a] The percentage of burned and unburned flint items is calculated within each lithic category

percentage of burning is recorded (3.13%). In addition, burning occurs on 0.09% of the flint pebbles (Table 3.12).

Burned flint microartifacts are distributed over most of the excavated area, with two areas displaying the highest frequencies of burning (Fig. 3.34). The first, in the center of the exposed surface, covers an area of 1.25 m^2 and includes 24.18% of the burned flint microartifacts. The other, slightly to the northwest, includes 21.93% of the burned flint microartifacts in a 1 m^2 area (Fig. 3.34). Comparison of these patterns with the distribution of unburned flint microartifacts demonstrates that the 1.25 m^2 of the central concentration of burned flint microartifacts display similarly higher relative percentages of unburned flint microartifacts (28.16%). A different pattern is evident in the 1 m^2 northwestern concentration, where the relative percentage of unburned flint microartifacts is significantly smaller (11.51%) than that of the burned ones (Fig. 3.34).

The density map of the burned flint microartifacts similarly emphasizes the northwestern concentration as exhibiting the highest density values, while the unburned flint microartifacts display two different areas of high density, in the center and in the southeastern corner of the excavated surface (Fig. 3.35).

The central area correspondingly displays high levels of density in all three random density maps and in two of these, high levels of density are evident in the southeastern corner as well (Fig. 3.36). The northwestern concentration, which displays the highest levels of density within the burned flint microartifacts and in which a substantial difference was observed between the relative percentages of the burned (21.93%) and unburned (11.51%) flint microartifacts, does not display high levels of density in any of the random density maps (Fig. 3.36).

The various distribution and density maps suggest that the distribution of the burned flint microartifacts is not uniform. The chi square test of the burned flint microartifacts further supports this observation ($\Sigma\chi^2 = 673.37$; $df = 65$; $p < 0.001$). The major contributor to this pattern is the northwestern concentration of burned flint microartifacts. This concentration is not congruent with the density patterns of the unburned flint microartifacts and is not represented in the random density maps as the area likely to display high density of burning. The contribution of this area to the observed patterning is supported by a comparison between the observed and expected percentages of burned flint microartifacts (Fig. 3.37).

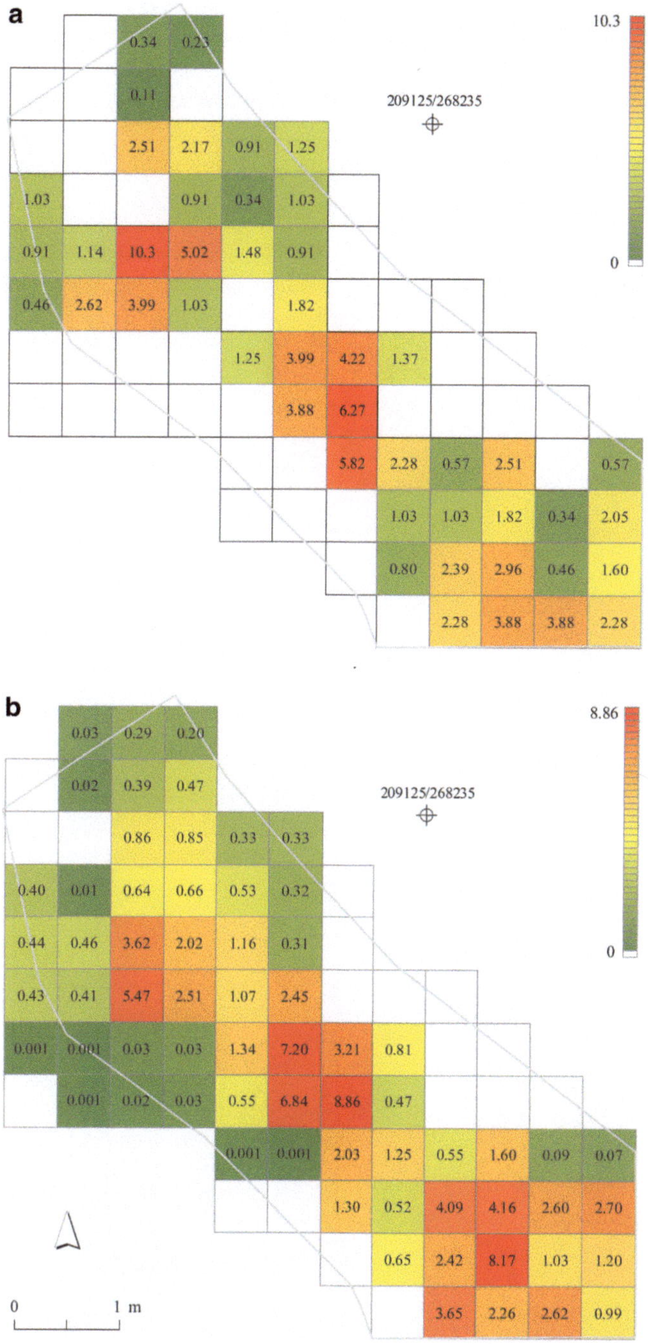

Fig. 3.34 Percentages of flint microartifacts per excavated unit in Layer II-6 L-3. (**a**) Burned flint microartifacts (N = 877) and (**b**) unburned flint microartifacts (N = 91,050)

Excavated units in which the observed values considerably exceed the expected ones occur in the vicinity of this concentration. Particularly, the sub-square associated with the kernel of the concentration, in which the percentage of burned flint microartifacts is 6.61% higher than the expected one, also

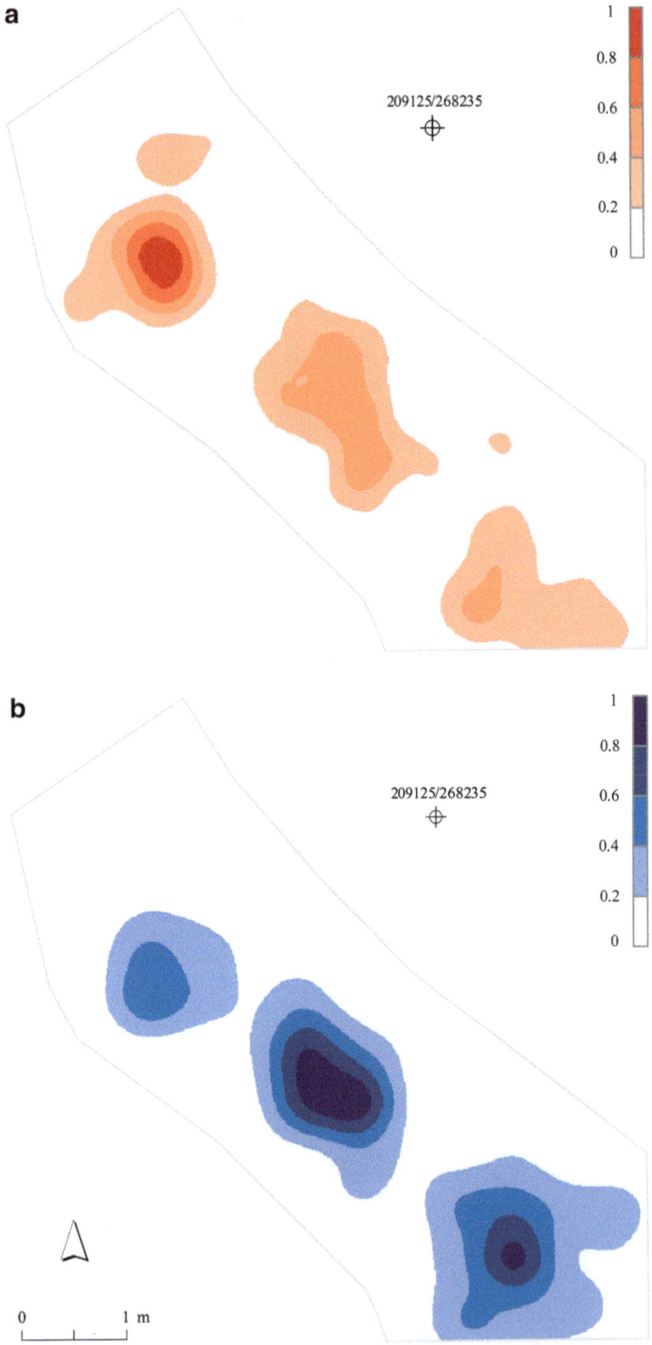

Fig. 3.35 Kernel density maps of flint microartifacts in Layer II-6 L-3. (**a**) Burned flint microartifacts (N = 877) and (**b**) unburned flint microartifacts (N = 91,050)

displays the highest SR value (SR = 10.18; N [expected] = 32.2). In the sub-square to the east, which covers the eastern edge of the kernel, the percentage of burned flint microartifacts is 2.96% higher than the expected one (SR = 6.16; N [expected] = 17.9). These point to this concentration as the key contributor to the observed clustering of the burned flint microartifacts.

A single burned flint pebble and 16 burned flint macroartifacts were recorded for Layer II-6 L-3 (Table 3.12). These seem

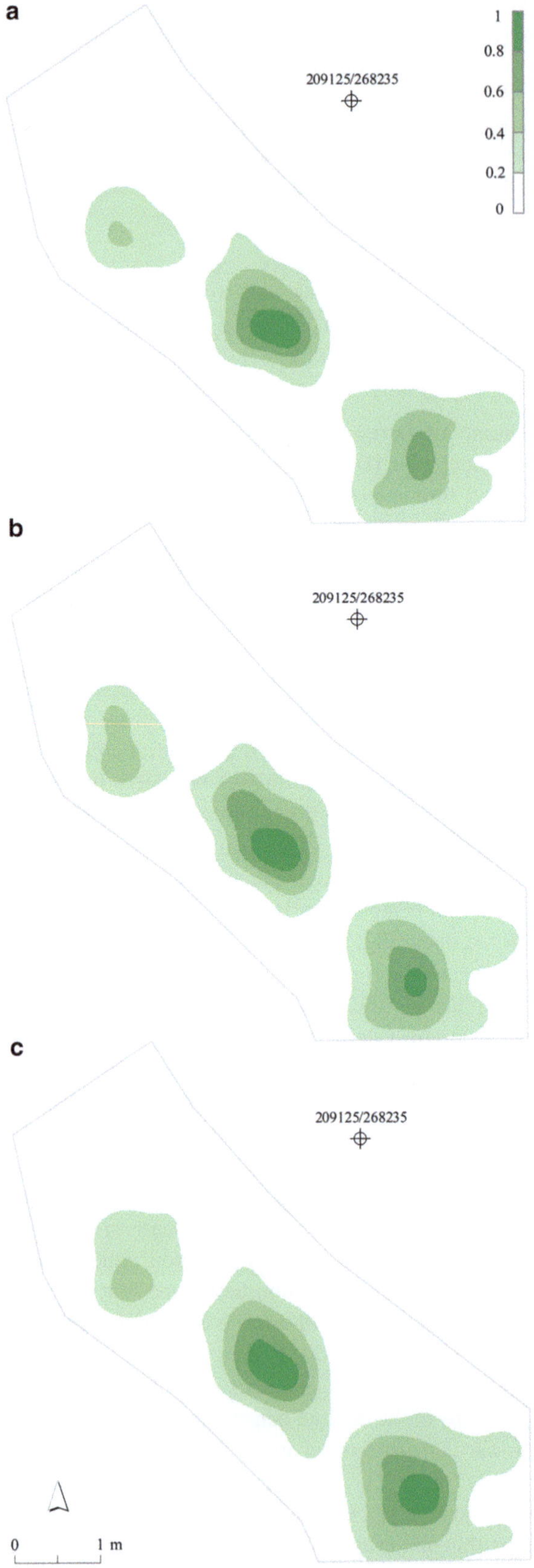

Fig. 3.36 Kernel density maps of three randomly selected data sets (N = 877) for Layer II-6 L-3

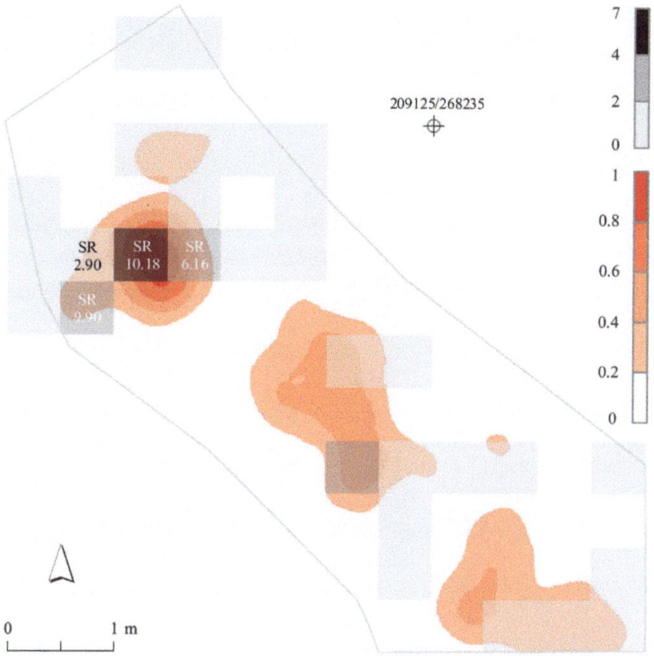

Fig. 3.37 The kernel density map of burned flint microartifacts of Layer II-6 L-3 and excavated units in which the observed percentage of burning exceeds the expected percentage; significant SR values are marked on the map

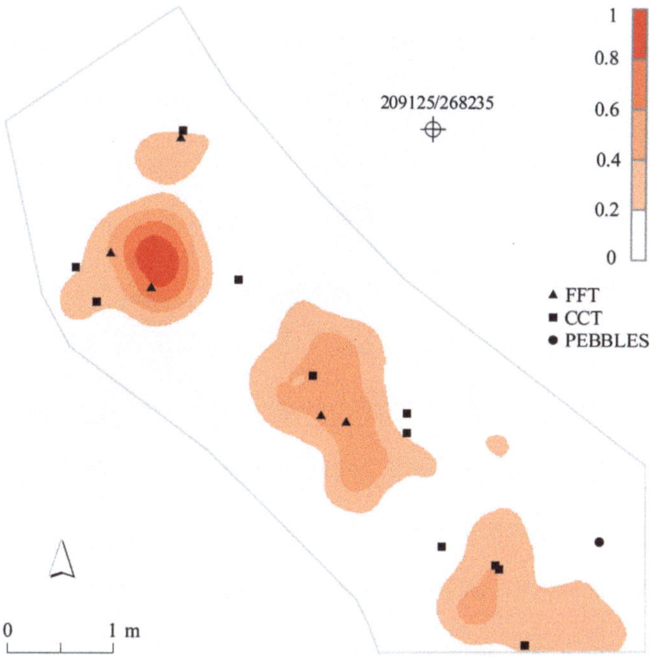

Fig. 3.38 The kernel density map of burned flint microartifacts of Layer II-6 L-3 and the distribution of large burned flint items (FFT: N = 5; CCT: N = 11; Pebbles: N = 1)

to be scattered throughout the excavated surface, and only four of the burned flint macroartifacts occur within the high-density concentration of burned flint microartifacts (Fig. 3.38).

3.11 Layer II-6 L-4

The 16.64 m^2 surface of Layer II-6 L-4 yielded the largest assemblage of microartifacts (N = 118,434) within the stratigraphic sequence of Gesher Benot Ya'aqov. While the majority of microartifacts are modified on flint, amongst the 1,729 macroartifacts basalt is the predominant raw material (Table 3.13). Burned items comprise 1.32% of the flint microartifacts, 2.45% of the flint macroartifacts, and only 0.85% of the flint pebbles (Table 3.13).

Burned flint microartifacts occur in most of the excavated units, in varying frequencies. The highest frequencies of burning are observed in two areas, in the center of the exposed surface, where four sub-squares incorporate 21.76% of the burned flint microartifacts, and in the southeastern part, with six sub-squares of high percentages adding up to 22.69% of the burned flint microartifacts (Fig. 3.39).

Unburned flint microartifacts display generally similar patterns, with the highest frequencies in the center and in the southeastern parts of the excavated surface (Fig. 3.39). The high frequencies of the central area are more dispersed and are observed within some ten sub-squares, differing from the more clustered pattern (i.e., four sub-squares of high frequency) observed in that area for the burned flint microartifacts. These four sub-squares, in which high frequencies of burned flint microartifacts are observed (a total of 21.76%), consist of 18.59% of the unburned flint microartifacts (Fig. 3.39).

A more substantial difference between the distribution of the burned and unburned flint microartifacts is observed in the southeastern area. There, the relative frequencies of the unburned flint microartifacts are higher than those of the burned ones. However, the six sub-squares in which the observed frequencies of burned flint microartifacts are high (adding up to 25.82%) contained only 17.03% of the unburned flint microartifacts (Fig. 3.39).

The general resemblance in the distribution of the burned and unburned flint microartifacts is illustrated in the density maps (Fig. 3.40). Both burned and unburned flint microartifacts

Table 3.13 Lithic assemblage of Layer II-6 L-4

Category	Flint				Basalt	Limestone	Total
	Unburned		Burned				
	N	%	N	%	N	N	N
Microartifacts[a]	105,022	98.67	1,406	1.32	7,279	4,727	118,434
FFT artifacts[a]	358	97.81	8	2.18	721	22	1,109
CCT artifacts[a]	275	97.17	8	2.82	95	23	401
Handaxes	2		–		159	–	161
Cleavers	–		–		58	–	58
Pebbles[a]	578	99.14	5	0.85	869	344	1,796
Total	106,235	98.67	1,427	1.32	9,181	5,116	121,959

[a]The percentage of burned and unburned flint items is calculated within each lithic category

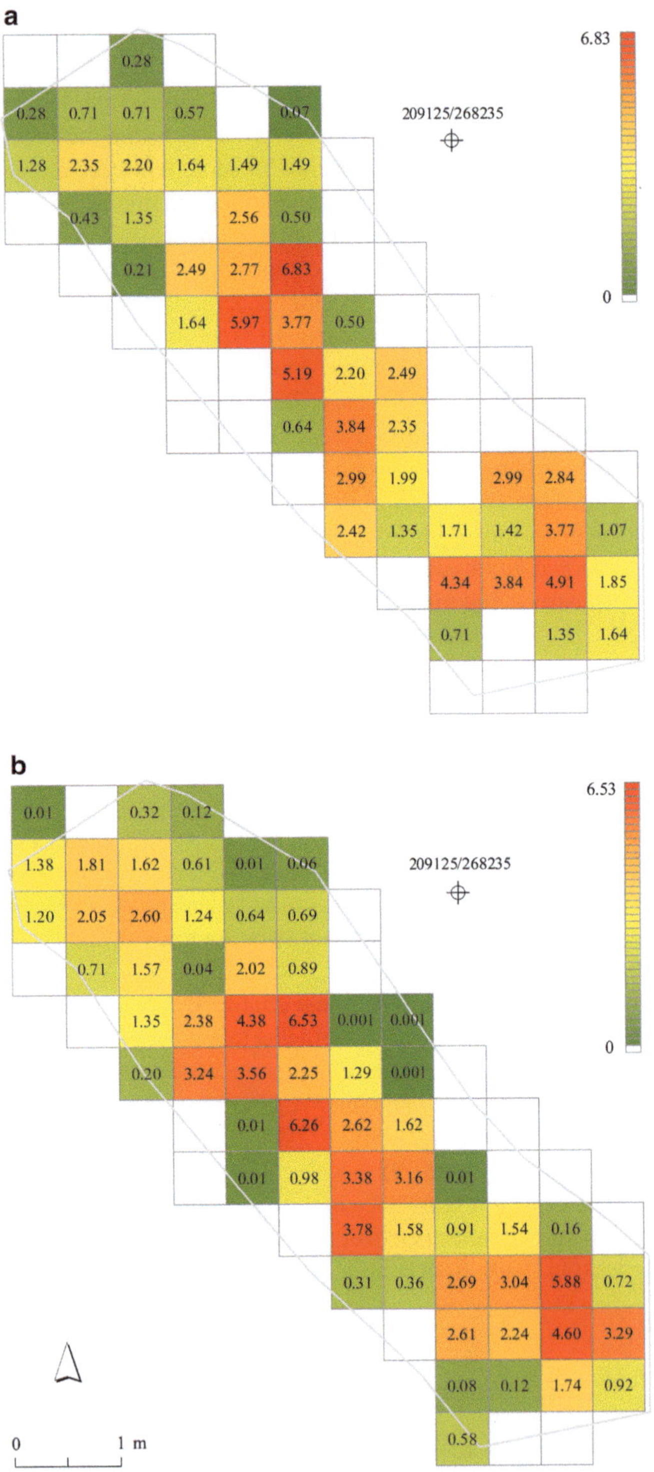

Fig. 3.39 Percentages of flint microartifacts per excavated unit in Layer II-6 L-4. (**a**) Burned flint microartifacts (N = 1,406) and (**b**) unburned flint microartifacts (N = 105,022)

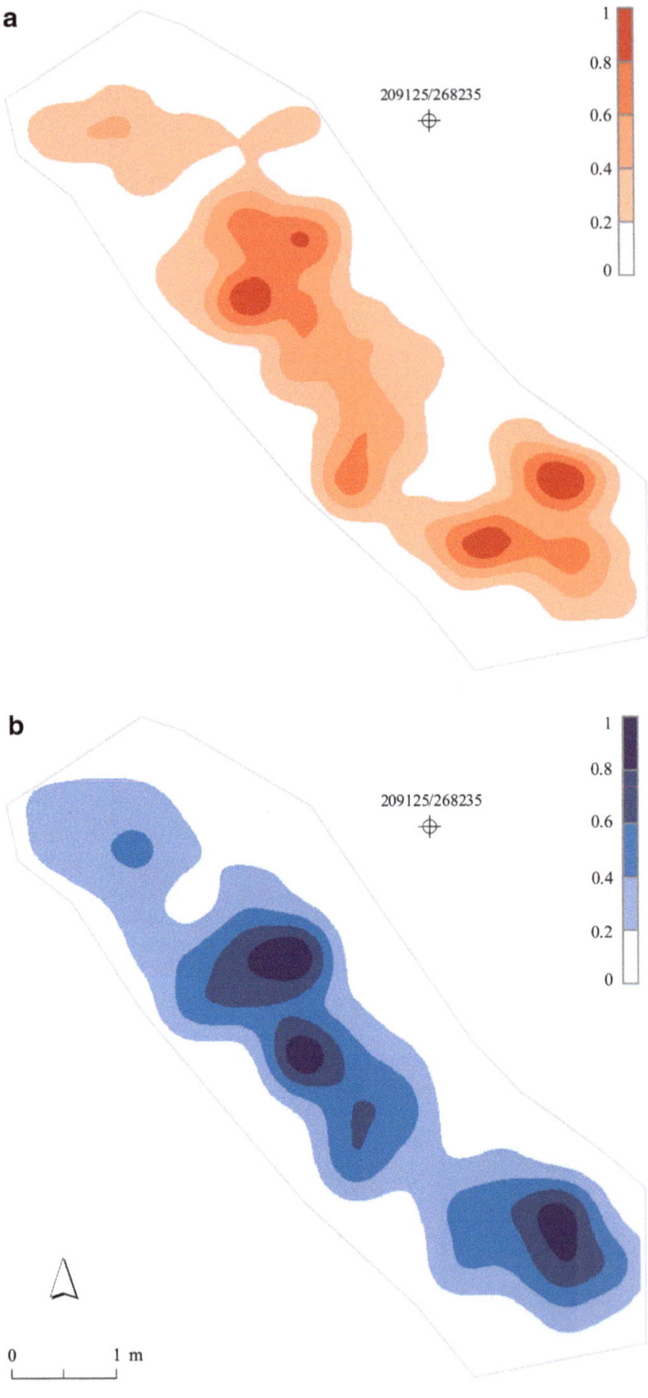

Fig. 3.40 Kernel density maps of flint microartifacts in Layer II-6 L-4. (**a**) Burned flint microartifacts (N = 1,406) and (**b**) unburned flint microartifacts (N = 105,022)

exhibit the highest levels of density in the center and in the southeastern area of the excavated surface. Due to the differences in relative frequencies of the burned and unburned flint microartifacts within particular excavated units (discussed above), the high-density clusters of the burned and unburned flint microartifacts exhibit different configurations and do not entirely overlap each other (Fig. 3.40).

The three random density maps (Fig. 3.41) exhibit greater similarities to the density patterns observed for the unburned flint microartifacts (Fig. 3.40). In particular, the configuration of the random dense clusters resembles those of the unburned

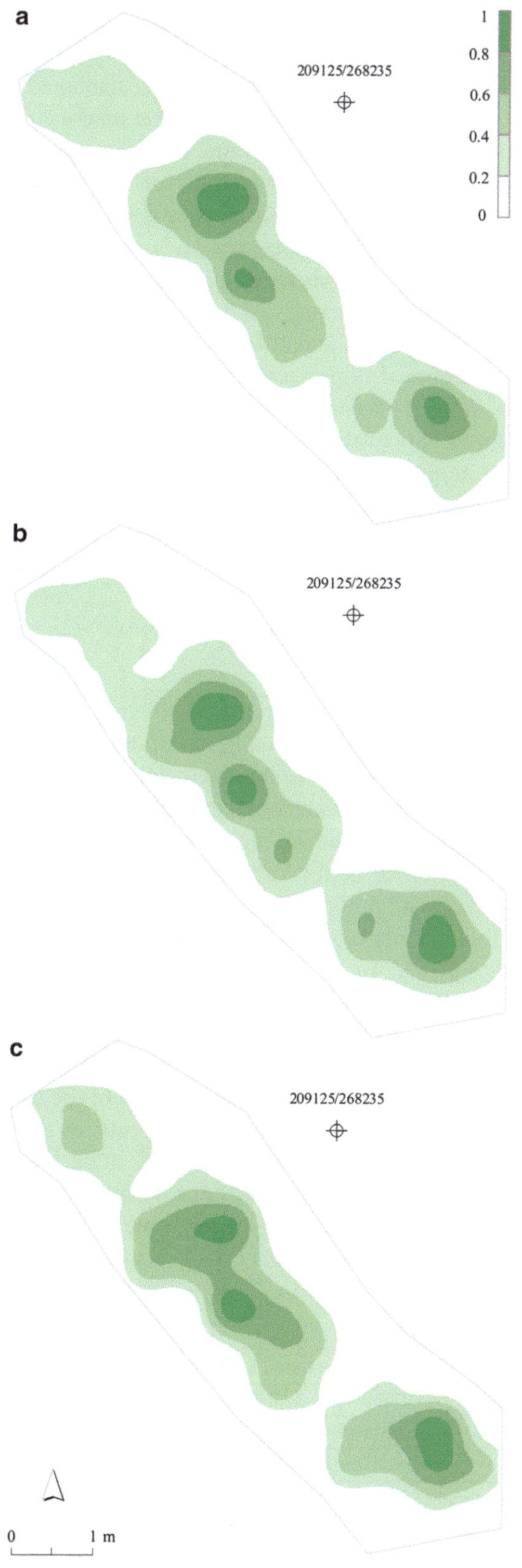

Fig. 3.41 Kernel density maps of three randomly selected data sets (N = 1,406) for Layer II-6 L-4

flint microartifacts more closely than those of the burned ones. Comparison of the density patterns of the burned flint microartifacts suggests that the observed difference is the result of the location of the kernels of the burned concentrations; the location of the kernel (i.e., the highest density level) of the observed burned flint microartifacts (Fig. 3.40) does not coincide with those of the random clusters (Fig. 3.41).

The various distribution and density maps suggest that the burned flint microartifacts are not evenly distributed. Rather, they display two areas in which density levels are relatively high. This patterning is confirmed by the chi square test applied to the burned flint microartifacts, which suggests that the distribution of these is significantly different from an expected, uniform distribution ($\Sigma\chi^2 = 1{,}063.87$; $df = 60$; $p < 0.001$). Calculation of standardized residuals for burned flint microartifacts of different excavated units suggested that significant values (SR > 2), pointing to the potential contributors of the observed clustering, occur in sub-squares within and in the vicinity of the burned concentrations, specified as follows.

The two areas in which the density and the frequencies of burning are high, occur in the southeastern part (henceforth: southern) and to its northwest (henceforth: northern) (Fig. 3.40).

Examination of the deviation between the observed percentages of burned flint microartifacts and the expected ones demonstrates that the major deviations, in which the observed percentage of burning exceeds the expected one, occur within the observed concentrations of high density of burning (Fig. 3.42). This pattern is characteristic of both the southern and the northern concentration. In the center of the exposed surface, a lower-density concentration is also associated with a higher (2.07%) percentage of observed burned flint microartifacts than what we would expect in a uniform distribution of the burning. However, since it is associated with a lower density concentration it is considered less significant.

In the northern concentration, several excavated units display higher percentages of burning than what we would expect if the distribution of the burned flint microartifacts were uniform (Fig. 3.42). Particularly, the sub-square that encircles the larger kernel of the concentration displays a 2.38% deviation between the observed and expected values of burning; the SR value for this sub-square is significant (SR = 4.71; N [expected] = 50.5). The sub-square next to it on the east, where the observed percentage of burning is 1.49% higher than expected, the SR value is significant as well (SR = 3.73; N [expected] = 31.9). The kernel of this concentration covers an area of 0.121 m^2 and includes 3.34% of the burned flint microartifacts and 1.79% of the unburned ones.

In the southern concentration, two kernels of high density value are observed. The first includes the sub-square in which the observed percentage of burned flint microartifacts is 2.64% higher than the expected one; the highest recorded SR

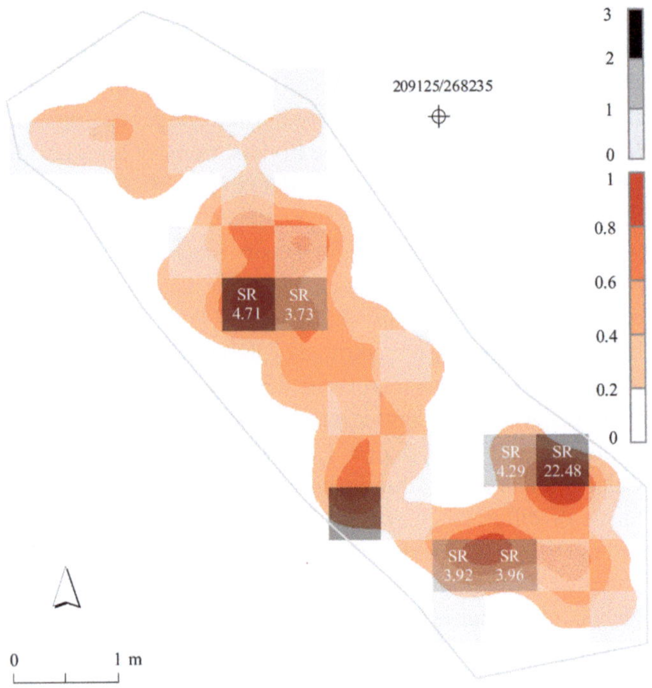

Fig. 3.42 The kernel density map of burned flint microartifacts of Layer II-6 L-4 and excavated units in which the observed percentage of burning exceeds the expected percentage; significant SR values are marked on the map

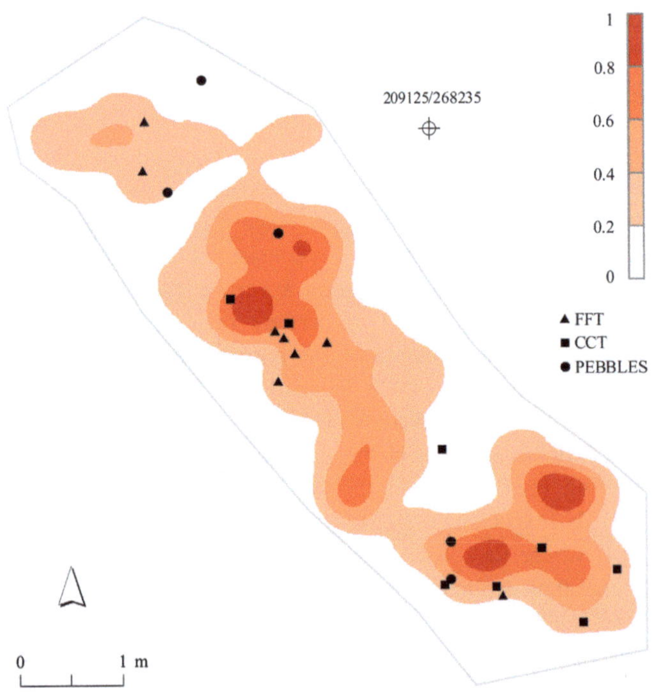

Fig. 3.43 The kernel density map of burned flint microartifacts of Layer II-6 L-4 and the distribution of large burned flint items (FFT: N = 8; CCT: N = 8; Pebbles: N = 5)

value is recorded in this sub-square (SR = 22.48; however N [expected] = 2.74). Within the sub-square next to it (to the west), the observed percentage is 1.42% higher than the expected one and the SR value is 4.29 (N [expected] = 21.9). This kernel covers an area of 0.137 m^2 and consists of 5.83% of the burned flint microartifacts and 1.67% of the unburned ones.

The second high-density kernel of the southern concentration is slightly southwest of the first and is encircled by two sub-squares in which the observed percentage is higher than the expected one, 1.58% higher in the eastern sub-square with an SR value of 3.96 (N [expected] = 31.7) and 1.69% higher in the western sub-square where SR = 3.92 (N [expected] = 37.1). Within this 0.111 m^2 kernel, 3.27% of the burned and 1.00% of the unburned flint microartifacts are included.

Sixteen burned flint macroartifacts and five burned flint pebbles were recovered from Layer II-6 L-4 (Table 3.13). The central part of the excavated surface is almost devoid of these items, and the majority of these occur in vicinity to the kernels of the southern (N = 6) and northern (N = 7) concentrations of burned flint microartifacts (Fig. 3.43).

3.12 Layer II-6 L-4b

The 13.69 m^2 exposure of Layer II-6 L-4b yielded some 8,778 microartifacts and 768 macroartifacts of different raw materials (Table 3.14). Basalt is extremely dominant amongst the macroartifacts and flint predominates amongst microartifacts.

The highest percentage of burned flint microartifacts within the Gesher Benot Ya'aqov sequence is observed in this layer. Some 5.80% of the flint microartifacts, 3.20% of the flint macroartifacts, and 1.94% of the flint pebbles are burned (Table 3.14).

Burned flint microartifacts are recorded in 31 of the 63 excavated sub-squares. The highest frequencies, adding up to 25.70% of the burned flint microartifacts, occur within two sub-squares in the southeastern corner of the excavated area. Relatively high frequencies also occur in the central part of the exposed surface (where 20.54% of the burned flint microartifacts occur in four sub-squares) and in the northern part, where three sub-squares incorporate together 16.02% of the burned flint microartifacts (Fig. 3.44).

Table 3.14 Lithic assemblage of Layer II-6 L-4b

Category	Flint				Basalt	Limestone	Total
	Unburned		Burned				
	N	%	N	%	N	N	N
Microartifacts[a]	7,182	94.19	443	5.80	1,008	145	8,778
FFT artifacts[a]	127	98.44	2	1.55	462	25	616
CCT artifacts[a]	21	87.50	3	12.50	42	6	72
Handaxes	3		–		57	2	62
Cleavers	–		–		18	–	18
Pebbles[a]	202	98.05	4	1.94	493	41	740
Total	7,535	94.34	452	5.65	2,080	219	10,286

[a] The percentage of burned and unburned flint items is calculated within each lithic category

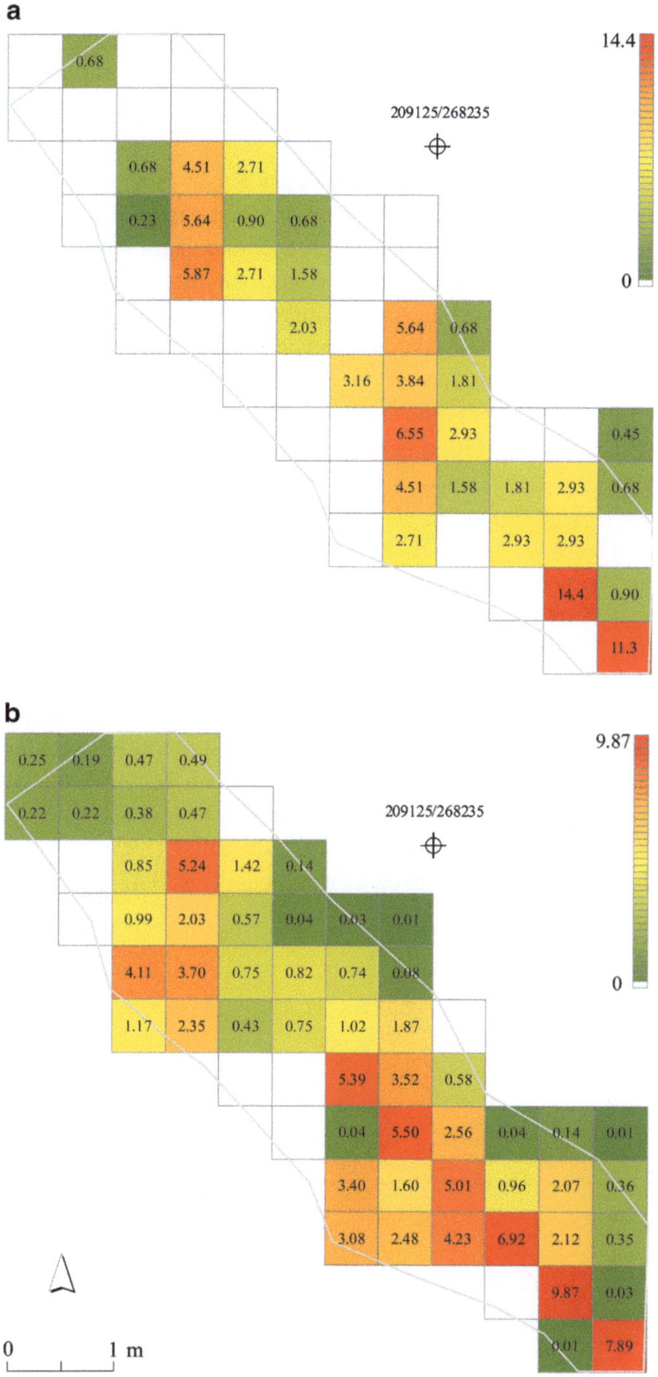

Fig. 3.44 Percentages of flint microartifacts per excavated unit in Layer II-6 L-4b. (**a**) Burned flint microartifacts (N = 443) and (**b**) unburned flint microartifacts (N = 7,182)

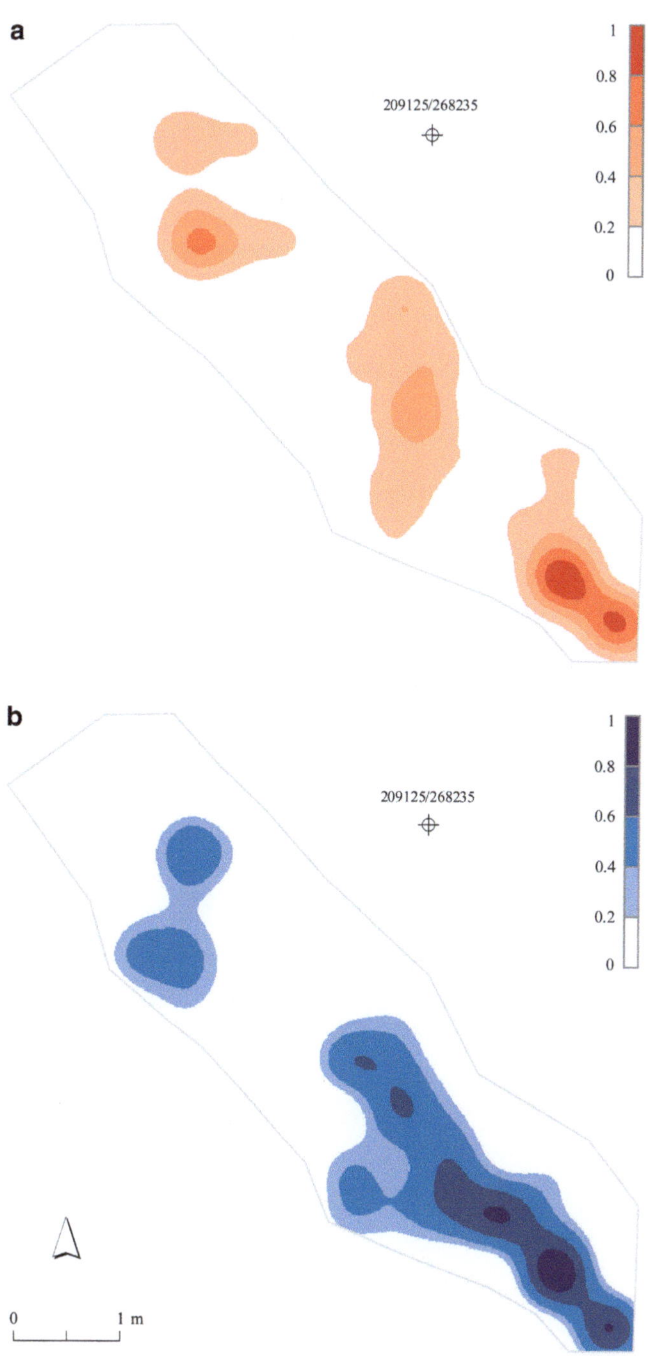

Fig. 3.45 Kernel density maps of flint microartifacts in Layer II-6 L-4b. (**a**) Burned flint microartifacts (N = 443) and (**b**) unburned flint microartifacts (N = 7,182)

Unburned flint microartifacts occupy a larger area, occurring in most of the excavated units. In comparison with the burned flint microartifacts, relatively high percentages of unburned flint microartifacts are recorded in a larger number of sub-squares. These occur in the southeastern, central, and northwestern parts of the surface, similarly to the general pattern of distribution of the burned flint microartifacts (Fig. 3.44).

A single concentration in which density levels are high is illustrated in the density map of the burned flint microartifacts in the southeastern corner of the exposed surface. The relatively high percentages observed in the central and northwestern areas (Fig. 3.44) are not depicted as high levels of density (Fig. 3.45).

Patterns of density within the unburned flint microartifacts similarly highlight the southeastern corner as displaying the highest levels of density; however, the dense cluster appears to be larger and more diffuse for the unburned flint microartifacts (Fig. 3.45).

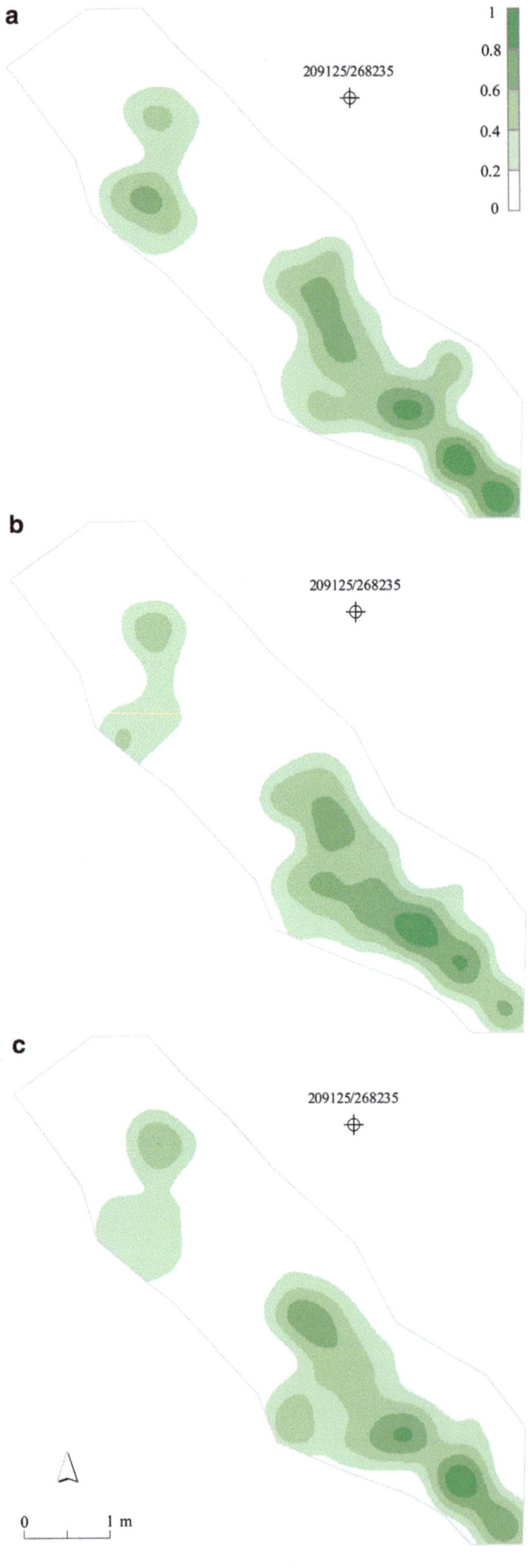

Fig. 3.46 Kernel density maps of three randomly selected data sets (N = 443) for Layer II-6 L-4b

A large and diffuse configuration is likewise illustrated in the random density maps. The southeastern area is highlighted in all three random density maps, and the general patterning resembles that observed for the unburned flint microartifacts (Fig. 3.46).

Thus, the distribution and density maps suggest a non-uniform distribution of the burned flint microartifacts and illustrate a single high-density concentration in the southeastern corner of the excavated surface. Correspondingly, the chi square test applied to the burned flint microartifacts suggests that the distribution of these is significantly different from an expected, uniform distribution ($\Sigma\chi^2 = 386.27$; $df = 55$; $p < 0.001$).

Differences between the observed and expected percentages of burning (specified below) in the area of the southeastern concentration suggest that this area is the key contributor to the observed pattern.

High density values are represented in this concentration in two kernels, which together encompass an area of 0.178 m^2 and incorporate 16.47% of the burned flint microartifacts of the layer and 5.82% of the unburned ones. The sub-square that displays the highest deviation between the observed and expected percentages of burning (4.30%) encircles a high-density kernel and exhibits a significant SR value (SR = 2.86; N [expected] = 44.8). In the sub-square that encircles the smaller high-density kernel, the observed percentage of burning is 3.20% higher than the expected one, and the SR value (SR = 2.37; N [expected] = 35.8) is likewise significant (Fig. 3.47).

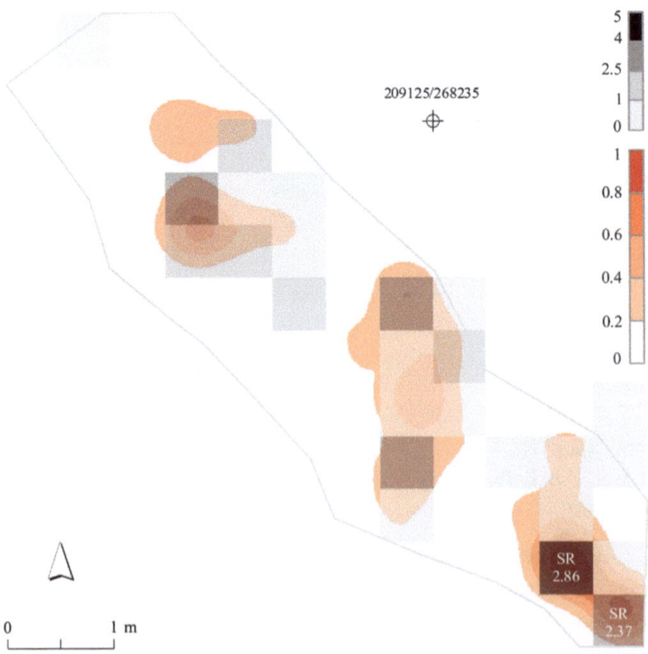

Fig. 3.47 The kernel density map of burned flint microartifacts of Layer II-6 L-4b and excavated units in which the observed percentage of burning exceeds the expected percentage; significant SR values are marked on the map

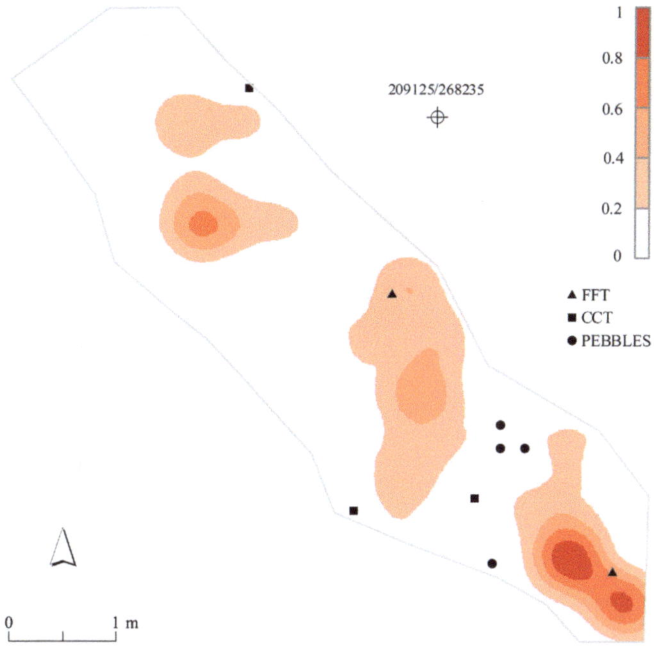

Fig. 3.48 The kernel density map of burned flint microartifacts of Layer II-6 L-4b and the distribution of large burned flint items (FFT: N = 2; CCT: N = 3; Pebbles: N = 4)

Five burned flint macroartifacts and four burned flint pebbles were recorded within the lithic assemblage of Layer II-6 L-4b (Table 3.14). One burned flint macroartifact occurs within the high-density concentration of burned flint microartifacts and several other burned items are located in close vicinity to it (Fig. 3.48). Three burned flint pebbles are recorded within a single sub-square some 1 m northwest of the high-density concentration of burned flint microartifacts (Fig. 3.48).

3.13 Layer II-6 L-5

The 13.39 m^2 of exposed surface in Layer II-6 L-5 uncovered some 37,609 microartifacts and 450 macroartifacts, modified predominantly on flint. Relatively high frequencies of burned flint are recorded for both microartifacts (3.66%) and macroartifacts (6.06%). In contrast, only 0.78% of the flint pebbles are burned (Table 3.15).

Table 3.15 Lithic assemblage of Layer II-6 L-5

Category	Flint				Basalt	Lime-stone	Total
	Unburned		Burned				
	N	%	N	%	N	N	N
Microartifacts[a]	31,862	96.33	1,211	3.66	2,540	1,996	37,609
FFT artifacts[a]	147	96.71	5	3.28	185	24	361
CCT artifacts[a]	37	84.09	7	15.90	15	1	60
Handaxes	2		–		14	–	16
Cleavers	–		–		13	–	13
Pebbles[a]	379	99.21	3	0.78	874	145	1,401
Total	32,427	96.35	1,226	3.64	3,641	2,166	39,460

[a] The percentage of burned and unburned flint items is calculated within each lithic category

Burned flint microartifacts are spread all over the exposed surface, occurring in most (48 out of 63) of the excavated sub-squares. High percentages are recorded for quite a few sub-squares, which extend from the center to the northern part of the exposed surface (Fig. 3.49). Apart from the two distinct sub-squares in the northern part, the distributions of high-frequency units seem relatively dispersed (Fig. 3.49). Comparable patterns emerge from the distribution map of the unburned flint microartifacts, which is largely similar to that of the burned flint microartifacts (Fig. 3.49).

This observed resemblance in distribution is further illustrated in the density maps of the burned and unburned flint microartifacts (Fig. 3.50). The two maps overlap almost completely; three dense areas are marked, with a similar location of the high-density kernels. Minor differences are observed within the configuration of the dense clusters: for the unburned flint microartifacts the central concentration appears more continuous, and midway between the central and northern concentrations the burned flint microartifacts display an additional, smaller, concentration that does not exhibit the high density level observed in the other concentrations (Fig. 3.50).

Correspondingly, the random density maps are notably similar to the density patterns observed within the flint microartifacts, both burned and unburned. On the whole, however, they display greater similarity to the density patterns of the unburned flint microartifacts, for example in a more continuous configuration of the central concentration (Fig. 3.51).

The distribution and density patterns of the burned flint microartifacts are very similar to those of the unburned ones. Neither is distributed evenly throughout the exposed surface and both exhibit higher frequencies in the central and northern parts of the excavated area. The chi square test results for the burned flint microartifacts suggest that their observed distribution is significantly different from an expected, uniform distribution ($\Sigma\chi^2 = 241.03$; $df = 54$; $p < 0.001$). However, calculations of standardized residuals for different excavated units imply that the major contributors to this are sub-squares that lie beyond the concentrations of burning (discussed below).

Examination of the difference between the observed and expected percentages of burning suggests generally minor variations; where the observed percentages of burned flint microartifacts are higher than the expected ones, they do not occur in the inner kernels of the concentrations and in no case exceed 2.00% (Fig. 3.52). For instance, close to the southern dense concentration, the two sub-squares to the south display higher observed percentages of burning (Fig. 3.52). These are recorded virtually outside the concentration and display relatively minor deviations between the observed and expected percentage of burned flint microartifacts; however, their SR values are significant (in the western sub-square: 1.31% deviation; SR = 2.55; N [expected] = 39.01;

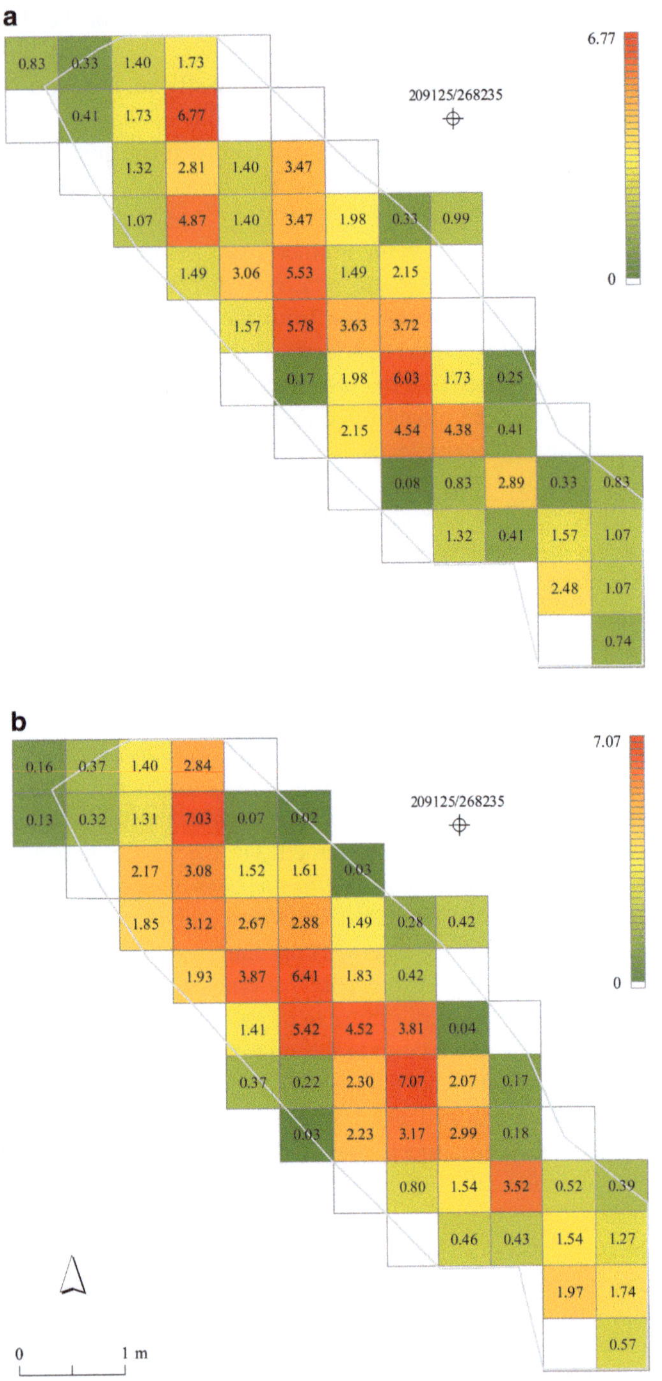

Fig. 3.49 Percentages of flint microartifacts per excavated unit in Layer II-6 L-5. (**a**) Burned flint microartifacts (N = 1,211) and (**b**) unburned flint microartifacts (N = 31,862)

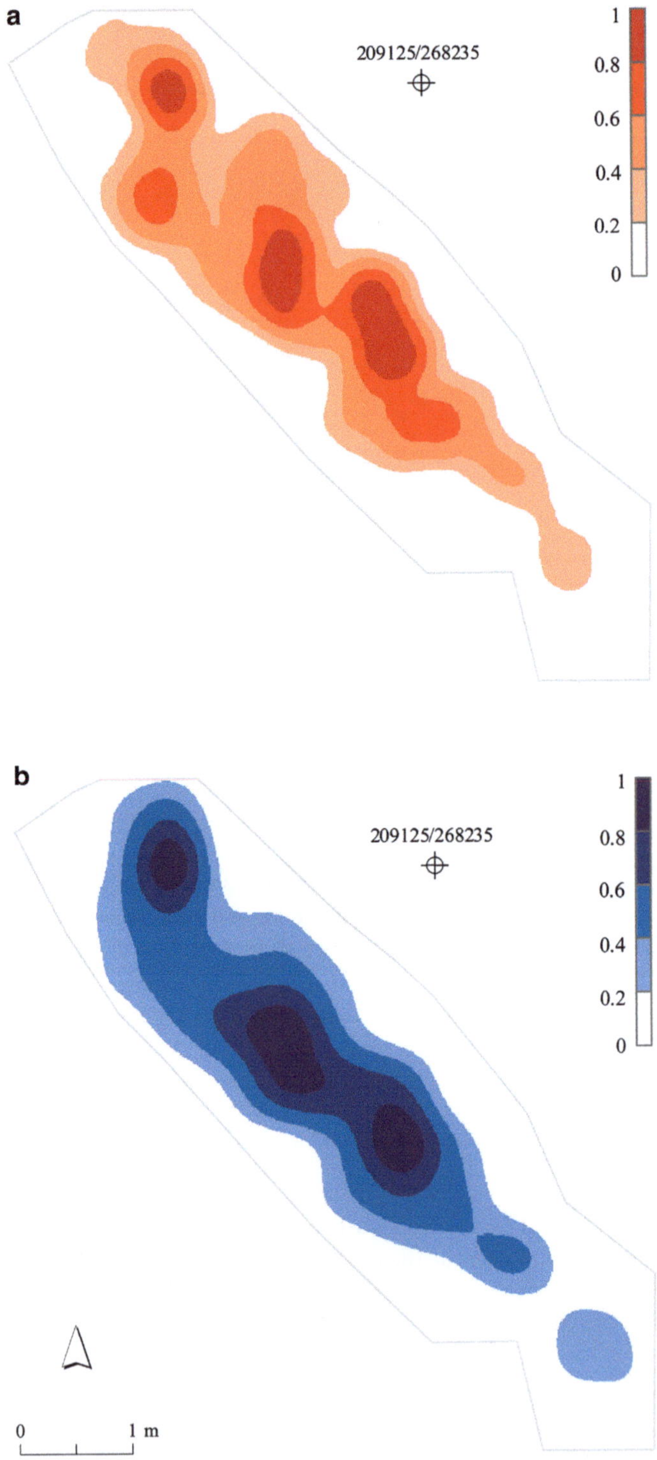

Fig. 3.50 Kernel density maps of flint microartifacts in Layer II-6 L-5. (**a**) Burned flint microartifacts (N = 1,211) and (**b**) unburned flint microartifacts (N = 31,862)

in the eastern sub-square: 1.33% deviation; SR = 2.65; N [expected] = 36.85). The highest recorded SR value is observed in a sub-square located on the northern edge of this concentration. This sub-square displays a 1.66% deviation between the observed and expected percentage of burning and SR = 8.36 (N [expected] = 5.81) (Fig. 3.52).

Overall, the burned flint microartifacts exhibit three high-density concentrations; two in the central part of the exposed surface (henceforth: southern and central) and another in the northern area (henceforth: northern).

The kernel of the southern concentration covers an area of 0.335 m^2 and includes 9.66% of the burned flint microartifacts and 6.44% of the unburned ones; the kernel of the central concentration covers an area of 0.227 m^2 and includes 6.52%

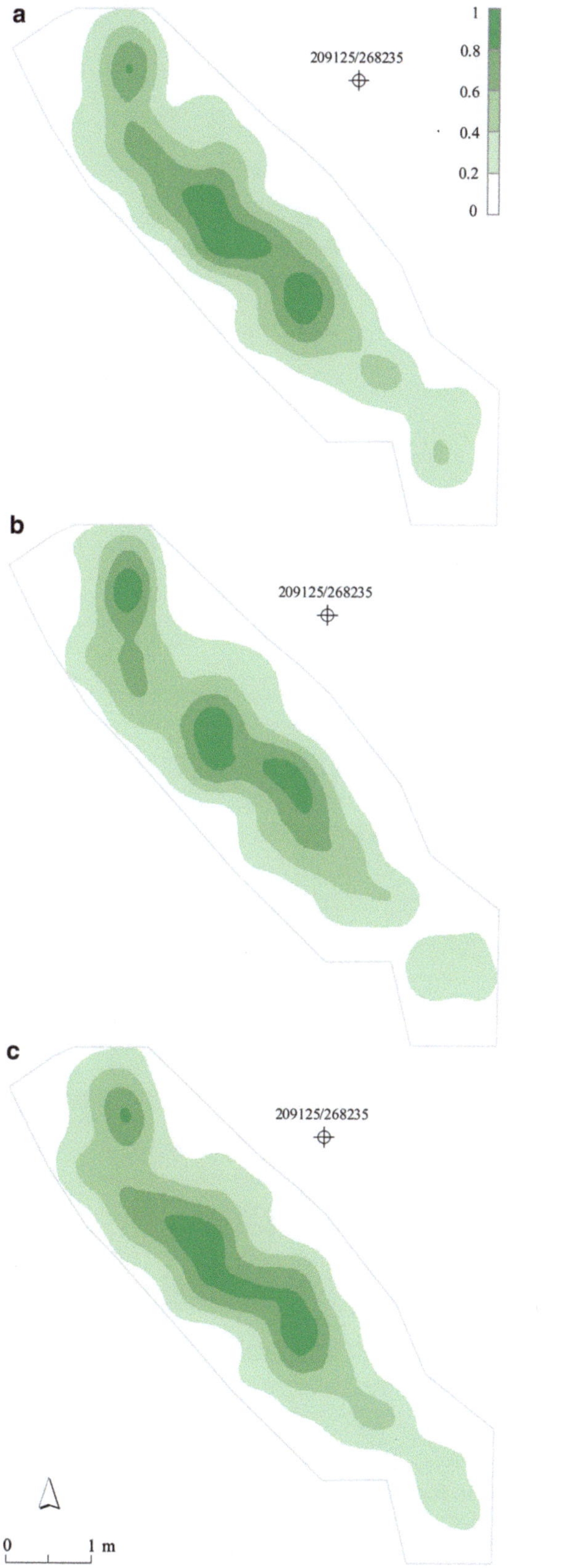

Fig. 3.51 Kernel density maps of three randomly selected data sets (N = 1,211) for Layer II-6 L-5

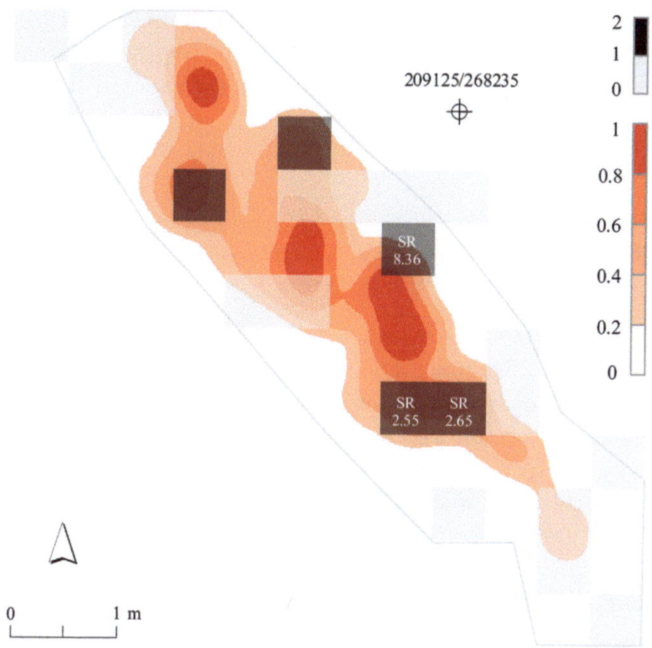

Fig. 3.52 The kernel density map of burned flint microartifacts of Layer II-6 L-5 and excavated units in which the observed percentage of burning exceeds the expected percentage; significant SR values are marked on the map

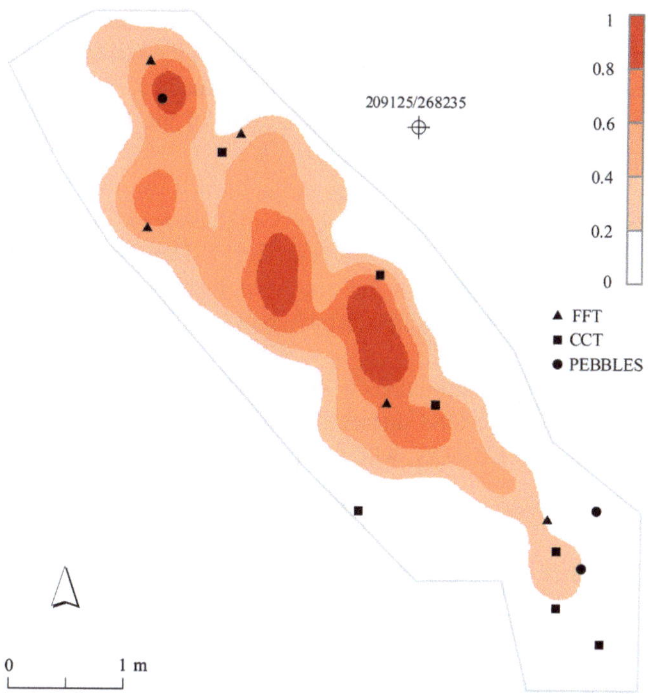

Fig. 3.53 The kernel density map of burned flint microartifacts of Layer II-6 L-5 and the distribution of large burned flint items (FFT: N = 5; CCT: N = 7; Pebbles: N = 3)

of the burned flint microartifacts and 5.42% of the unburned ones; and the kernel of the northern concentration covers an area of 0.095 m^2 and includes 4.12% of the burned flint microartifacts and 2.59% of the unburned ones.

In addition to burned flint microartifacts, Layer II-6 L-5 yielded 12 burned flint macroartifacts and three burned flint pebbles (Table 3.15). Six of these occur within a relatively small area (ca. 1.5 m²) in the southeastern part of the excavated area, while the remaining nine burned items seem to be scattered throughout the exposed surface (Fig. 3.53).

3.14 Layer II-6 L-6

The 12.62 m² of excavated surface in Layer II-6 L-6 yielded some 13,357 microartifacts and 732 macroartifacts (Table 3.16). Burning was observed on 2.00% of the flint microartifacts, on 2.69% of the flint macroartifacts, and on only 0.64% of the flint pebbles (Table 3.16).

Burned flint microartifacts are recorded in 36 of the 60 excavated sub-squares, with only a few of these exhibiting high percentages of burning. The highest percentages of burning occur in a 1.25 m² area (i.e., five sub-squares) in the northwestern corner of the excavated area, where they add up to 25.10% of the burned flint microartifacts, and in the southeastern corner, where an equivalent area contributes 26.69% of the burned flint microartifacts (Fig. 3.54).

Unburned flint microartifacts occupy a larger area and cover most of the exposed surface. The highest percentages of these occur in quite a few sub-squares in the southeastern, central, and northwestern parts of the surface, displaying a more dispersed distribution than that of the burned flint microartifacts (Fig. 3.54).

In terms of density levels, the unburned flint microartifacts show higher densities in the southeastern and northwestern parts of the surface (Fig. 3.55). The northwestern concentration overlaps the single high-density cluster of burned flint microartifacts, which, however, covers a smaller area (Fig. 3.55).

The three random density maps point to the southeastern part of the excavated surface as being most likely to display high levels of density, while the northwestern concentration is illustrated as an isolated high-density cluster in only one of the three random density maps (Fig. 3.56).

The distribution and density maps suggest a general overlapping between the burned and unburned flint microartifacts.

Table 3.16 Lithic assemblage of Layer II-6 L-6

Category	Flint				Basalt	Lime-stone	Total
	Unburned		Burned				
	N	%	N	%	N	N	N
Microartifacts[a]	12,282	97.99	251	2.00	493	331	13,357
FFT artifacts[a]	269	97.46	7	2.53	296	29	601
CCT artifacts	56	96.55	2	3.44	46	5	109
Handaxes	–		–		16	–	16
Cleavers	–		–		5	1	6
Pebbles[a]	463	99.35	3	0.64	2,231	218	2,915
Total	13,070	98.03	263	1.97	3,087	584	17,004

[a] The percentage of burned and unburned flint items is calculated within each lithic category

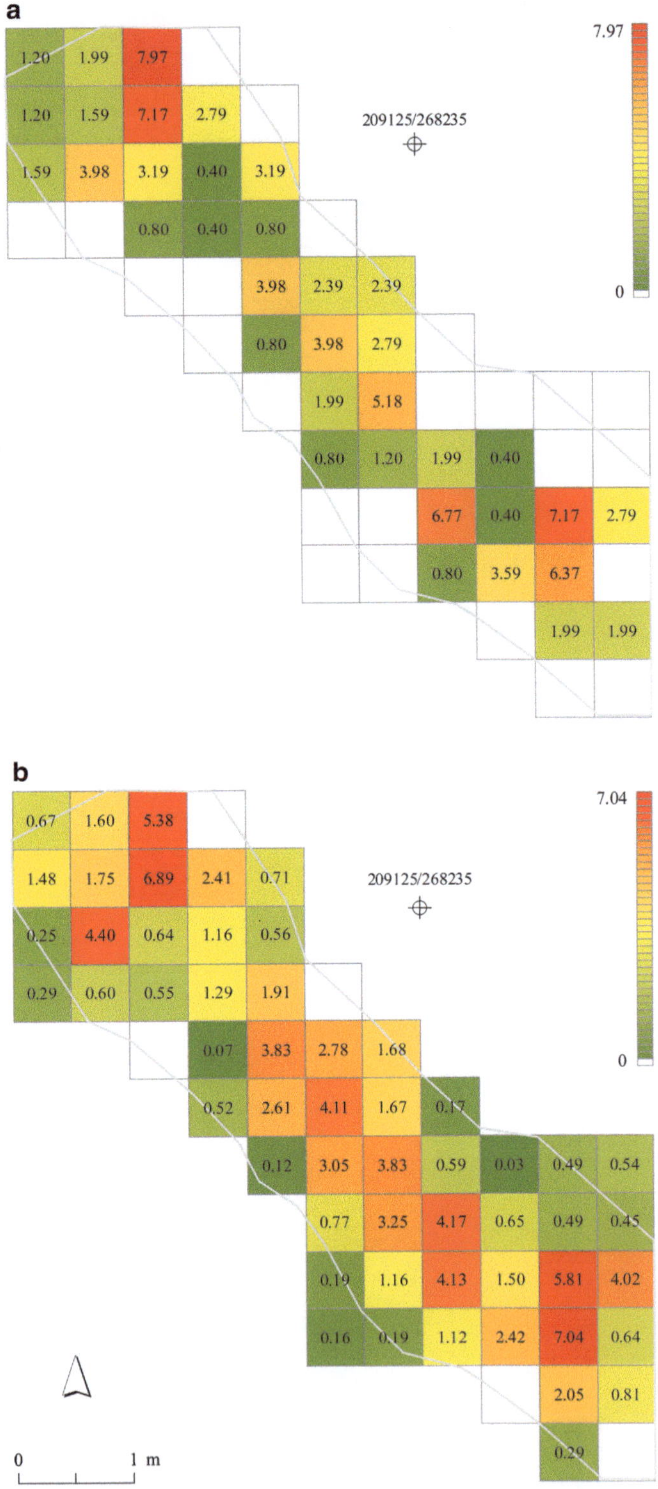

Fig. 3.54 Percentages of flint microartifacts per excavated unit in Layer II-6 L-6. (**a**) Burned flint microartifacts (N = 251) and (**b**) unburned flint microartifacts (N = 12,282)

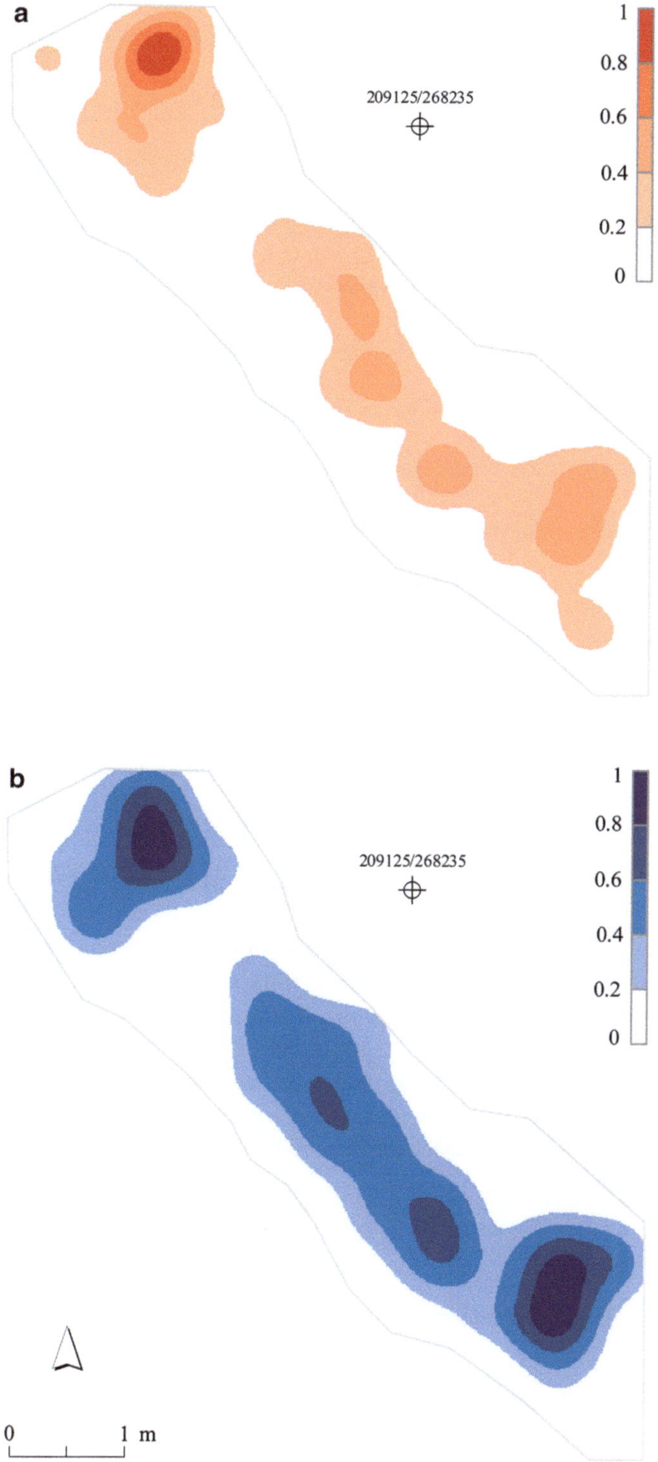

Fig. 3.55 Kernel density maps of flint microartifacts in Layer II-6 L-6. (**a**) Burned flint microartifacts (N = 251) and (**b**) unburned flint microartifacts (N = 12,282)

The northwestern concentration is, however, more marked within the distribution of the burned flint microartifacts, which exhibit a smaller, denser, concentration in this area. The chi square test applied to the burned flint microartifacts suggests that the distribution of these is significantly different

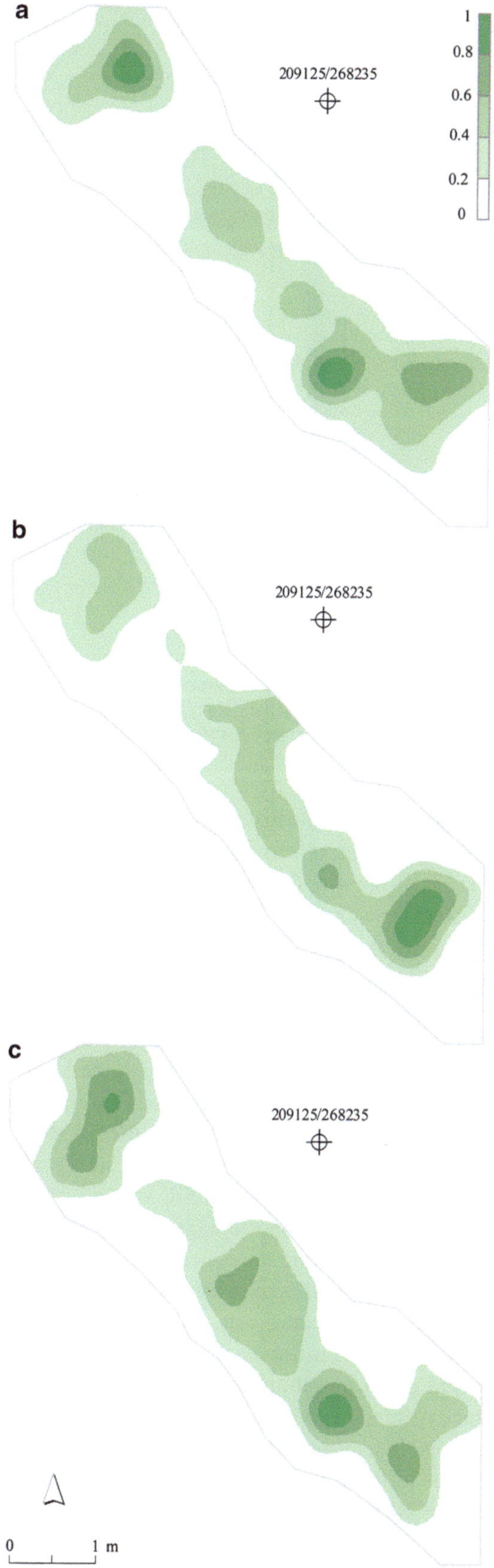

Fig. 3.56 Kernel density maps of three randomly selected data sets (N = 251) for Layer II-6 L-6

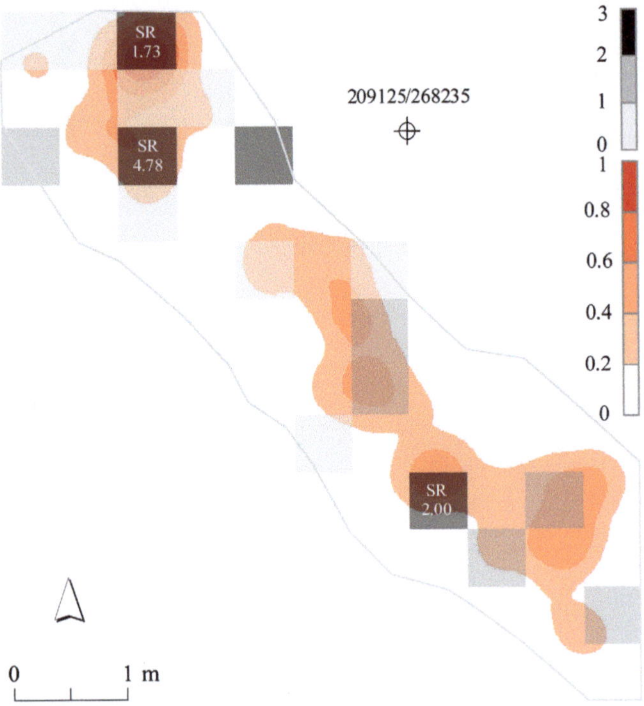

Fig. 3.57 The kernel density map of burned flint microartifacts of Layer II-6 L-6 and excavated units in which the observed percentage of burning exceeds the expected percentage; SR values are marked on the map

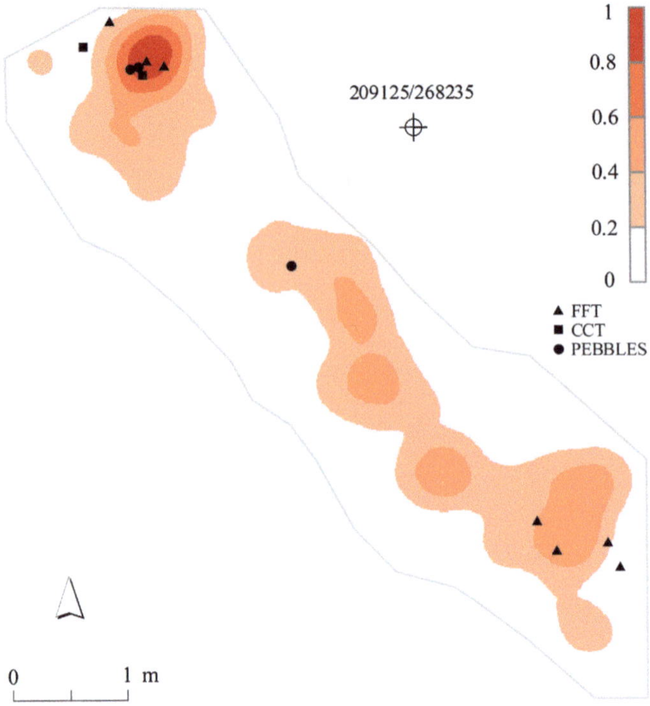

Fig. 3.58 The kernel density map of burned flint microartifacts of Layer II-6 L-6 and the distribution of large burned flint items (FFT: N = 7; CCT: N = 2; Pebbles: N = 3)

from an expected, uniform distribution ($\Sigma\chi^2 = 288.06$; $df = 57$; $p < 0.001$). Differences between the observed and expected percentages of burning (specified below) in the area of the northwestern concentration suggest that this area is the key contributor to the observed pattern.

Excavated units in which the observed percentages of burned flint microartifacts exceed the expected ones occur in the vicinity of the northwestern concentration (Fig. 3.57).

Within the sub-square associated with the kernel of the concentration, the percentage of burned flint microartifacts is 2.53% higher than the expected one; the SR value of this sub-square is, however, not significant (SR = 1.73; N [expected] = 13.6). Significant SR values are observed within two sub-squares on the edges of the concentration; in the first, located within the concentration on its southern edge, the percentage of burned flint microartifacts is 2.50% higher than the expected one (SR = 4.78; however N [expected] = 1.72); in the second, 0.5 m to the east, the percentage of burned flint microartifacts is 2.57% higher than the expected one (SR = 5.20; however N [expected] = 2.57). These excavated units may point to this area as a contributor to the observed differentiation in the distribution of the burned flint microartifacts. Another potential contributor is a sub-square located in the southeastern part of the exposed surface, where the observed percentage of burned flint microartifacts is 2.58% higher than the expected one (Fig. 3.57; the significance of the SR value for this sub-square is, however, not certain (SR = 2.00; N [expected] = 10.5).

The kernel of the northern concentration covers an area of 0.120 m^2 and comprises 7.96% of the burned flint microartifacts and 2.80% of the unburned ones.

The 12 burned flint items larger than 2 cm of Layer II-6 L-6 include nine macroartifacts and three pebbles (Table 3.16). Seven of these occur within and in close vicinity to the northern concentration of burned flint microartifacts (Fig. 3.58).

3.15 Layer II-6 L-7

3.15.1 Layer II-6 L-7: Upper Occupational Floor

The excavated surface of the upper occupational floor of Layer II-6 L-7 (12.60 m^2) yielded some 25,915 microartifacts and 1,098 artifacts, modified predominantly on flint (Table 3.17). Evidence of burning is seen on 2.80% of the flint microartifacts, 3.07% of the flint macroartifacts, and 0.63% of the flint pebbles (Table 3.17).

Burned flint microartifacts are recorded in most excavated units, covering 48 of the 62 excavated sub-squares. High per-

Table 3.17 Lithic assemblage of Layer II-6 L-7 (upper occupational floor)

Category	Flint				Basalt	Limestone	Total
	Unburned		Burned				
	N	%	N	%	N	N	N
Microartifacts[a]	23,455	97.19	677	2.80	955	828	25,915
FFT artifacts[a]	511	97.14	15	2.85	358	63	947
CCT artifacts[a]	87	95.60	4	4.39	37	6	134
Handaxes	–		–		8	–	8
Cleavers	–		–		9	–	9
Pebbles[a]	784	99.36	5	0.63	2,228	258	3,275
Total	24,837	97.25	701	2.74	3,595	1,155	30,288

[a] The percentage of burned and unburned flint items is calculated within each lithic category

centages of these are observed in several sub-squares, extending from the southern to the northern corner of the excavated area (Fig. 3.59). In the southeastern part of the exposed surface, three sub-squares (0.75 m^2) of relatively high percentages incorporate 16.40% of the burned flint microartifacts of the layer; slightly to the northwest, 12.84% are recorded within an equivalent area of 0.75 m^2; the remaining ten sub-squares in which relatively high percentages of burning are recorded extend from the center to the northwest, adding up to 41.79% of the burned flint microartifacts (Fig. 3.59).

The distribution of the unburned flint microartifacts displays similar patterns (Fig. 3.59). The three southeastern sub-squares in which the percentages of burning are relatively high similarly display relatively high frequencies and include 11.06% of the unburned flint microartifacts. The remaining excavated units in which the percentage of the unburned flint microartifacts is high encompass a larger area than that observed for the burned flint microartifacts; here, sixteen sub-squares of relatively high frequencies extend continuously from the center to the northwest and amount to 64.70% of the unburned flint microartifacts of the layer (Fig. 3.59).

Thus both the burned and unburned flint microartifacts display a generally similar pattern, with high percentages of microartifacts in a relatively continuous distribution extending from the center to the northwest (more markedly for the unburned flint microartifacts) and a single, more isolated area of high frequencies in the southeast. The percentage of the burned flint microartifacts within this southeastern concentration (16.40%) is higher than that of the unburned ones (11.06%).

These patterns are clearly illustrated in the density maps of the burned and unburned flint microartifacts (Fig. 3.60). The bulk of the exposed surface, from the center to the northwest, displays large concentrations of flint microartifacts with higher levels of density within the unburned than the burned ones. The southeastern concentration displays higher levels of density for the burned flint microartifacts (Fig. 3.60).

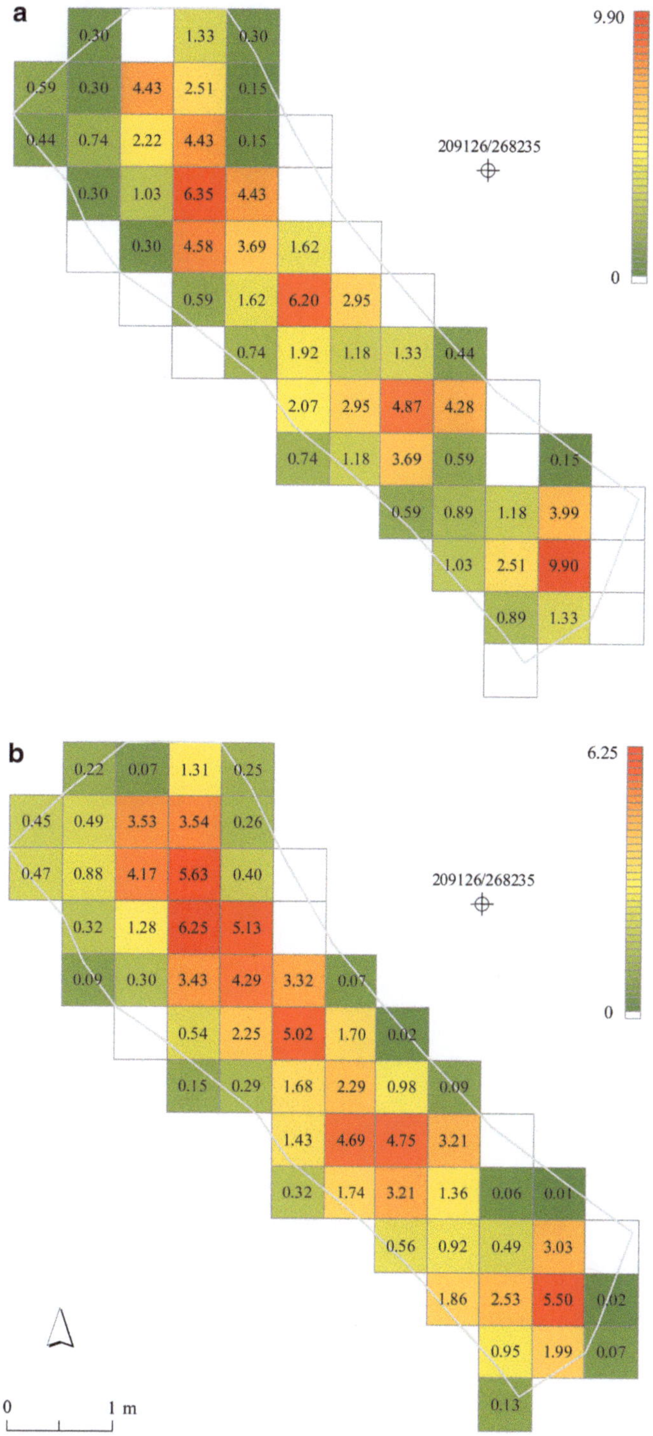

Fig. 3.59 Percentages of flint microartifacts per excavated unit in Layer II-6 L-7 upper occupational floor. (**a**) Burned flint microartifacts (N = 677) and (**b**) unburned flint microartifacts (N = 23,455)

The chi square test results for the burned flint microartifacts suggest that the distribution of these is significantly different from an expected, uniform distribution ($\Sigma\chi^2$ = 109.32; $df = 57$; $p < 0.001$). Calculation of standardized residuals for burned flint microartifacts of different excavated units

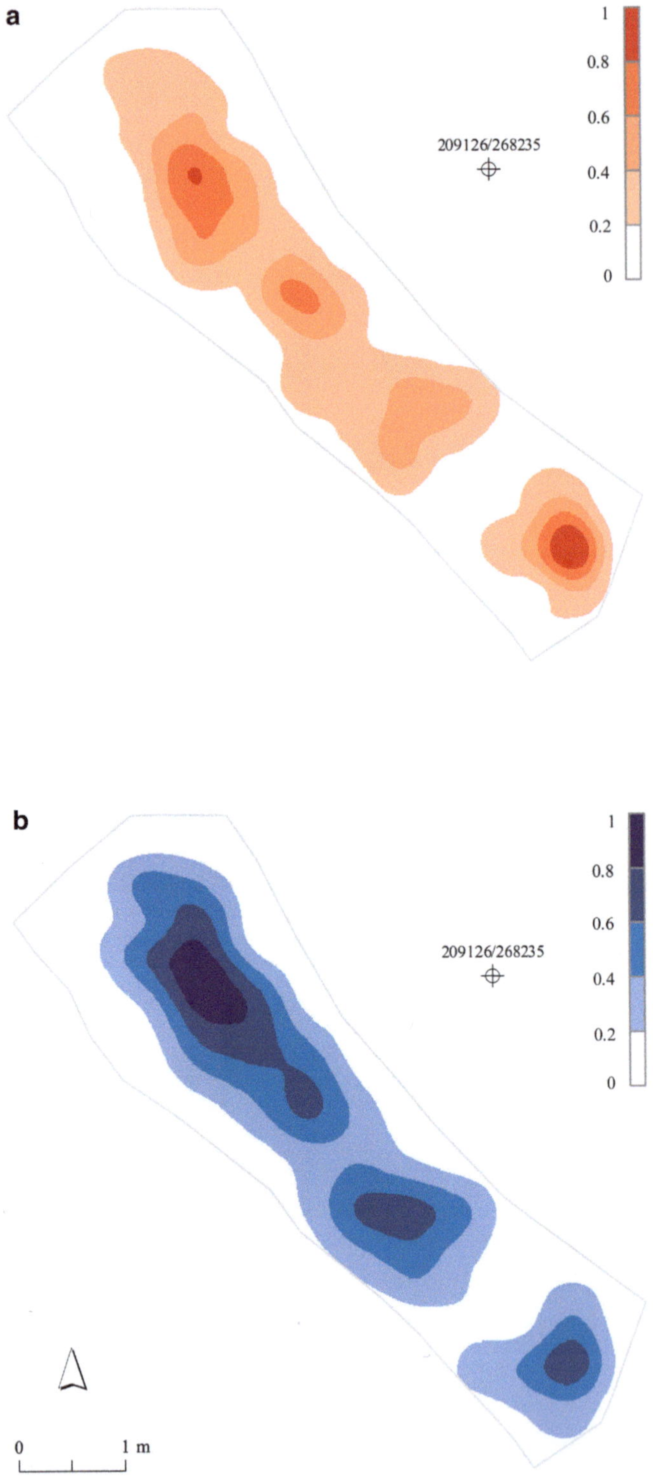

Fig. 3.60 Kernel density maps of flint microartifacts in Layer II-6 L-7 upper occupational floor. (**a**) Burned flint microartifacts (N = 677) and (**b**) unburned flint microartifacts (N = 23,455)

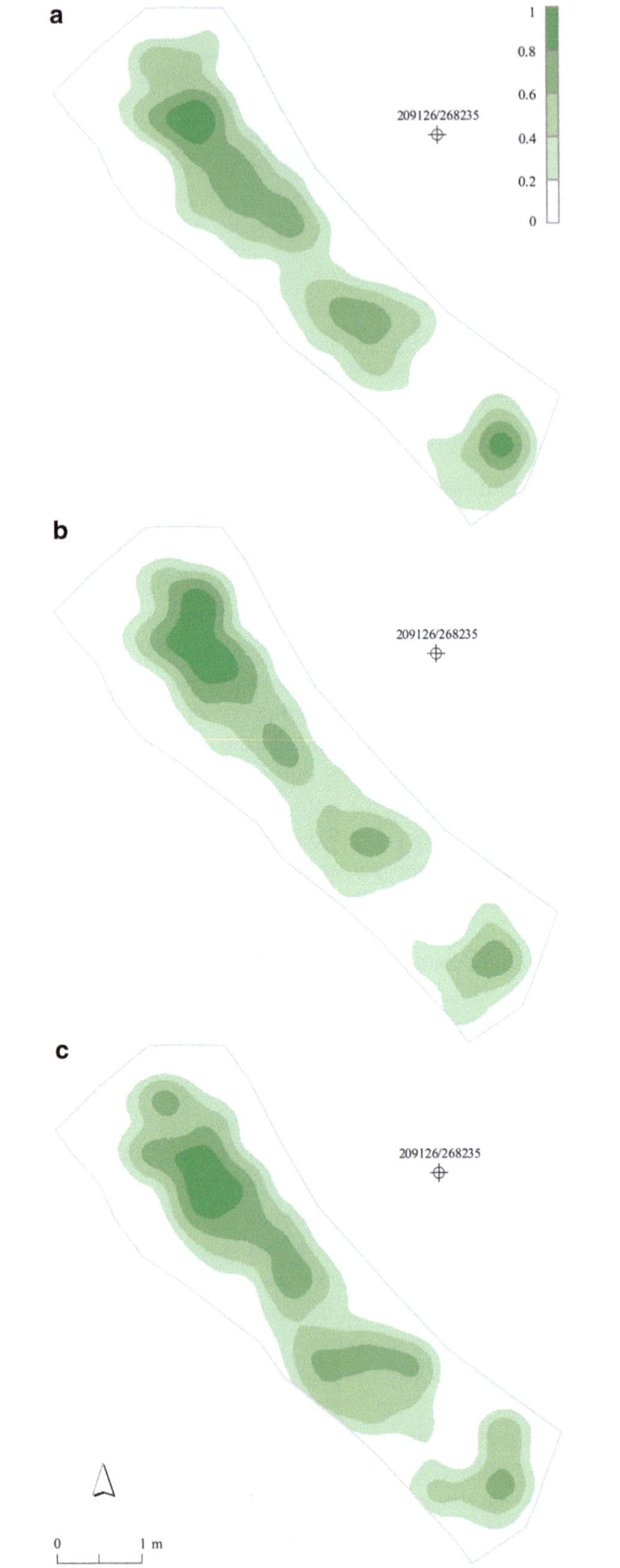

Fig. 3.61 Kernel density maps of three randomly selected data sets (N = 677) for Layer II-6 L-7 upper occupational floor

suggests that the southeastern concentration of the burned flint microartifacts is the major contributor to this pattern (discussed below).

The southeastern area appears in all three random density maps, reaching the fifth and highest level of density in only one of the three. In general, however, the observed patterns of distribution and density of the burned flint microartifacts resemble the patterns illustrated in the random density maps (Fig. 3.61).

Examination of the differences between the observed and expected percentages of burning within various excavated units reveals minor variations throughout the exposed surface. It is only within the southeastern concentration that the observed percentage of burned flint microartifacts notably exceeds the expected one (Fig. 3.62). Within the sub-square covering most of the kernel of this concentration, the observed percentage of burning is 4.27% higher than the expected one, with a significant SR value (SR = 4.70; N [expected] = 38). Merely for the sake of comparison, in the northern part of the exposed surface the burned flint microartifacts exhibit a concentration in which the high-density kernel is particularly small. Within the sub-square encircling this kernel, the percentage of burned flint microartifacts is only 0.09% higher than the expected one and the SR value is insignificant (SR = 0.10; N [expected] = 42.3) (Fig. 3.62).

Thus, the pattern observed within the southeastern concentration of burned flint microartifacts does not coincide with that of the unburned ones. Furthermore, the kernel of this concentration, which covers an area of 0.125 m^2, includes 8.12% of the burned and only 3.00% of the unburned flint microartifacts.

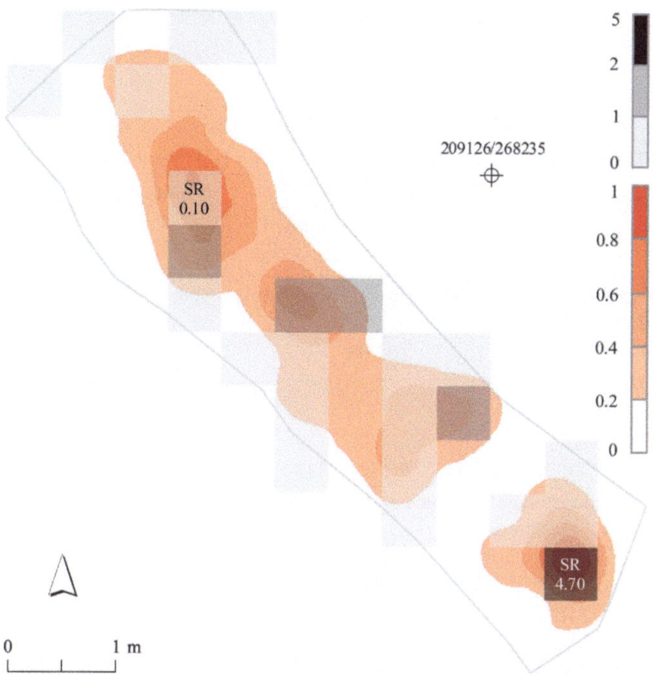

Fig. 3.62 The kernel density map of burned flint microartifacts of Layer II-6 L-7 upper occupational floor and excavated units in which the observed percentage of burning exceeds the expected percentage; SR values are marked on the map

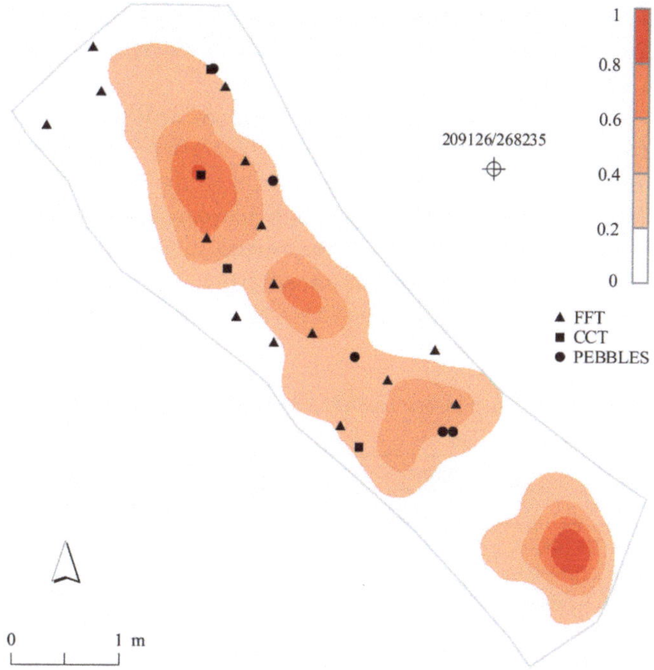

Fig. 3.63 The kernel density map of burned flint microartifacts of Layer II-6 L-7 upper occupational floor and the distribution of large burned flint items (FFT: N = 15; CCT: N = 4; Pebbles: N = 5)

The upper occupational floor of Layer II-6 L-7 yielded 19 burned flint macroartifacts and five burned flint pebbles (Table 3.17). None of these occurs in the vicinity of the southeastern concentration of burned flint microartifacts (Fig. 3.63).

3.15.2 Layer II-6 L-7: Test Pits

Following the removal of the assemblages of the upper occupational horizon of Layer II-6 L-7, two test pits (northern and southern) were excavated to a deeper extent, reaching the contact between II-6 and the underlying II-7 at the base of the level (the excavation of these test pits and the methodology that facilitated their spatial plotting are discussed in detail in Chapter 2, Section 2.4.4). The minimal spatial extent of these exposures does not permit a meaningful spatial analysis; thus, the following discussions present the general lithic inventory of these two test pits alongside schematic illustrations of the spatial distribution of the burned and unburned flint microartifacts.

3.15.2.1 The Northern Pit

The Northern Pit of Layer II-6 L-7 yielded some 12,555 microartifacts and 332 macroartifacts from a 2.75 m^2 area.

Flint is the dominant raw material, exhibiting burning on 2.62% of the microartifacts and 2.80% of the macroartifacts (Table 3.18).

The burned and unburned flint microartifacts appear to be distributed similarly and to overlap each other. Burned flint microartifacts occur in 8 of the 11 excavated units and display relatively high frequencies in several sub-squares; thus 86.4% of the burned flint microartifacts are recorded within four sub-squares (Fig. 3.64). These four sub-squares also incorporate 87.10% of the unburned flint microartifacts, which generally occur within all excavated units (Fig. 3.64).

Five burned macroartifacts were recovered from the Northern Pit of Layer II-6 L-7 (Table 3.18). These occur within two sub-squares in which relatively high frequencies of burned flint microartifacts are recorded (Fig. 3.64).

Table 3.18 Lithic assemblage of Layer II-6 L-7 (Northern Pit)

Category	Flint				Basalt	Lime-stone	Total
	Unburned		Burned				
	N	%	N	%	N	N	N
Microartifacts[a]	10,940	97.37	295	2.62	763	557	12,555
FFT artifacts[a]	155	97.48	4	2.51	119	13	291
CCT artifacts[a]	18	94.73	1	5.26	13	6	38
Handaxes	–		–		1	1	2
Cleavers	–		–		1	–	1
Pebbles	88	100.00	–		506	15	609
Total	11,201	97.39	300	2.60	1,403	592	13,496

[a] The percentage of burned and unburned flint items is calculated within each lithic category

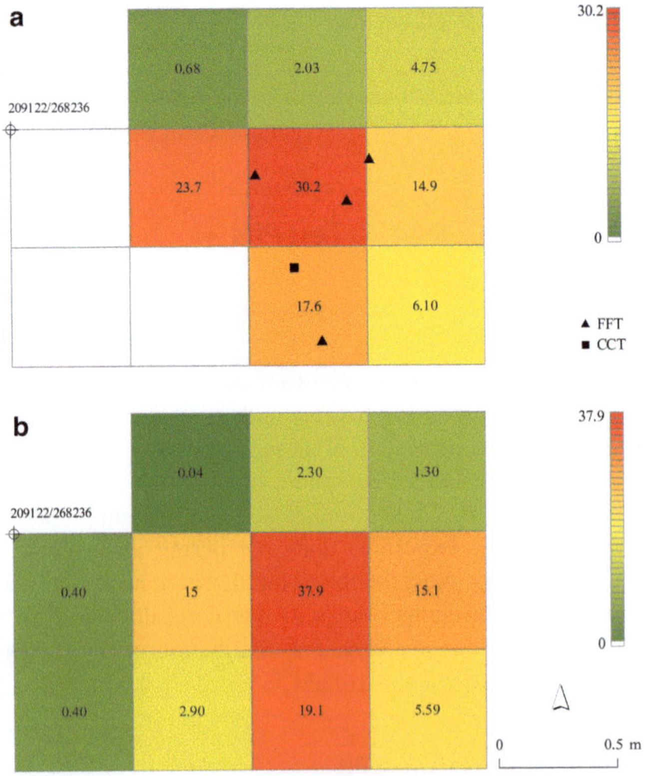

Fig. 3.64 Percentages of flint microartifacts per excavated unit in Layer II-6 L-7 Northern Pit. (**a**) Burned flint microartifacts (N = 295); distribution of burned flint macroartifacts is marked on the map (FFT: N = 4; CCT: N = 1) and (**b**) unburned flint microartifacts (N = 10,940)

3.15.2.2 The Southern Pit

The southern pit, covering an area of 4.25 m^2, yielded 6,874 microartifacts and 104 macroartifacts of various raw materials, predominantly flint (Table 3.19).

Evidence of burning was observed on 3.72% of the flint microartifacts and on 1.78% of the flint macroartifacts.

The distribution of the burned and unburned flint microartifacts exhibits a general overlapping. Of the 14 excavated units in which burned flint microartifacts occur, high percentages are recorded in eight; these excavated units correspond with the units of relatively high percentages of unburned flint microartifacts, with the exception of one sub-square in the western part of the exposed surface (Fig. 3.65).

In addition to the burned flint microartifacts, a single burned flint macroartifact, found in the central part of the excavated area (Fig. 3.65), was recorded in the Southern Pit of Layer II-6 L-7 (Table 3.19).

Table 3.19 Lithic assemblage of Layer II-6 L-7 (Southern Pit)

Category	Flint				Basalt	Lime-stone	Total
	Unburned		Burned				
	N	%	N	%	N	N	N
Microartifacts[a]	6,228	96.27	241	3.72	188	217	6,874
FFT artifacts[a]	40	97.56	1	2.43	35	3	79
CCT artifacts[a]	15	100.00	–		9	1	25
Handaxes	–		–		–	–	–
Cleavers	–		–		–	–	–
Pebbles	61	100.00	–		51	7	119
Total	6,344	96.32	242	3.67	283	228	7,097

[a] The percentage of burned and unburned flint items is calculated within each lithic category

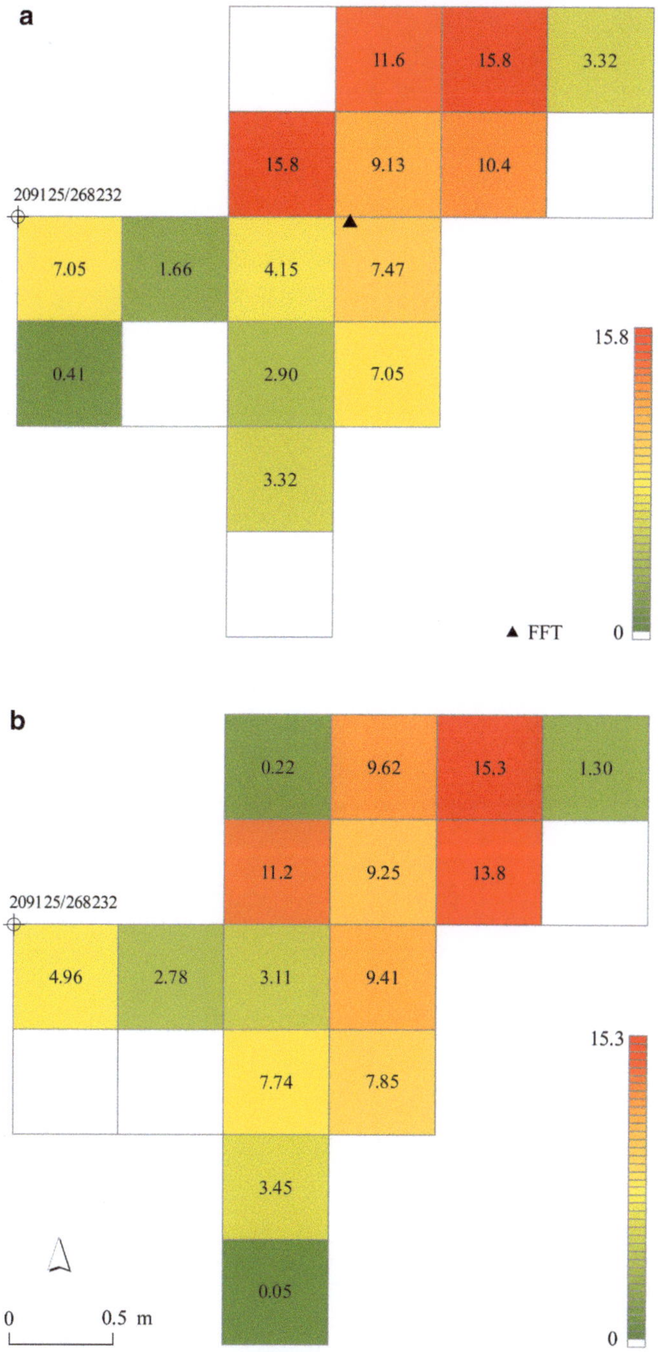

Fig. 3.65 Percentages of flint microartifacts per excavated unit in Layer II-6 L-7 Southern Pit. (**a**) Burned flint microartifacts (N = 241), location of burned flint macroartifact is marked on the map (FFT: N = 1) and (**b**) unburned flint microartifacts (N = 6,228)

Chapter 4
Discussion and Conclusions

The previous chapter presented data on the presence and spatial distribution of burned flint items throughout the stratigraphic sequence at GBY. It demonstrated that, while burned flint occurs in all the studied archaeological horizons, the burned flint microartifacts are not evenly distributed throughout the excavated surface and denser concentrations of burned material are observed (Table 4.1).

This chapter will investigate the possible agents influencing the spatial configuration of the burned flint items in the archaeological layers at GBY. In outlining the research hypotheses of this study (Section 2.3), three agents were considered as potentially affecting the spatial arrangement of the archaeological material, and specifically that of the burned flint items. These include lake-margin processes, fires controlled by humans, and natural fires. These issues are discussed in this chapter – the issue of natural fires (Section 4.1) and anthropogenic fires (Section 4.2). This is followed by an examination of the use of fire in relation to other aspects of hominin activities and behavior, as documented by different archaeological layers (Section 4.3). The conclusions emerging from this examination, and from this study in general, are incorporated in the concluding part of this chapter (Section 4.5).

The outline of the hypotheses of this study addressed the issue of lake-margin processes that may have affected the distribution patterns of small lithic artifacts. Review of this issue suggested that if the archaeological occurrences of GBY were subjected to such lake-margin processes, the expectation is that flint microartifacts would be distributed in particular configurations. This study has demonstrated that these configurations do not occur within the analyzed archaeological layers. The flint microartifacts do not exhibit a linear distribution along a presumed shoreline; rather, clusters of microartifacts are recorded and are always associated with larger artifacts (see details below). Thus, our basic hypothesis, which assumes the preservation of the original spatial configuration of the archaeological remains at GBY, is supported by the observed distribution patterns of the flint microartifacts and macroartifacts.

The spatial association between microartifacts and other components of the archaeological layers is presented in this chapter. Chapter 3 presented the data on the spatial configuration of the larger-sized burned flint items (i.e., macroartifacts and pebbles). With the exception of two layers (Layer V-5 and Layer II-6 L-7, upper occupational floor), in which the larger burned flint items are not spatially associated with clusters of burned flint microartifacts, the archaeological layers exhibit a spatial association between larger and smaller burned flints to various degrees (Table 4.2).

Table 4.2 and the distribution maps of the large burned flint items (presented in Chapter 3) demonstrate that in several layers most of the large burned items occur within or in close vicinity to the concentrations of burned flint microartifacts (i.e., Layers II-5, II-5/6, II-6 L-4, II-6 L-6). In other layers only a few burned large flints are associated with the clusters of burned flint microartifacts, while the remaining majority is scattered throughout the excavated surface (i.e., Layers II-6 L-1, II-6 L-2, II-6 L-3, II-6 L-4b, II-6 L-5).

The relatively small numbers of burned flint macroartifacts and pebbles do not provide sufficient sample sizes for further in-depth spatial analyses. However, their presence is significant for reasons beyond the spatial association of small and large items discussed above. These items will be referred to in subsequent parts of this chapter in the discussion of the issue of natural fires.

The results of this study have demonstrated that, although burned flint is present throughout the archaeological sequence, the relative frequencies, and particularly the patterns of distribution, vary between the different layers. This research has examined three models that may explain the presence and distribution of the flint items (see Section 2.3.3). *Model 1* suggested that an absence of burned flint items indicates that neither an anthropogenic nor a natural fire operated at the site. The results of this study have determined that we can confidently reject *Model 1*, as burned flint items occur in each of the examined archaeological layers. The other two models integrate the presence of burned flint, considering the degree of spatial association between

N. Alperson-Afil and N. Goren-Inbar, *The Acheulian Site of Gesher Benot Ya'aqov Volume II: Ancient Flames and Controlled Use of Fire*, Vertebrate Paleobiology and Paleoanthropology, DOI 10.1007/978-90-481-3765-7_4, © Springer Science+Business Media B.V. 2010

Table 4.1 Summary table – characteristics of the high-density kernels of burned clusters; measurements of length (L) and width (W) axis are given in meters; areas in m^2

Layer	Unburned flint microartifacts	Burned flint microartifacts	Area	L	W
II-5	N = 267 (7.73%)	N = 17 (10.49%)	0.219	0.61	0.51
II-5/6 North	N = 390 (4.59%)	N = 32 (8.00%)	0.167	0.97	0.23
II-5/6 South	N = 169 (1.99%)	N = 18 (4.5%)	0.122	0.40	0.38
II-6 L-1 North	N = 664 (1.25%)	N = 30 (3.97%)	0.148	0.46	0.41
II-6 L-1 South	N = 318 (0.59%)	N = 28 (3.71%)	0.117	0.46	0.31
II-6 L-2	N = 1,751 (2.39%)	N = 53 (9.41%)	0.165	0.51	0.40
II-6 L-3	N = 1,947 (2.13%)	N = 59 (6.72%)	0.161	0.50	0.41
II-6 L-4 North	N = 1,884 (1.79%)	N = 47 (3.34%)	0.121	0.41	0.37
II-6 L-4 Southeast	N = 1,764 (1.67%)	N = 82 (5.83%)	0.137	0.47	0.36
II-6 L-4 Southwest	N = 1,054 (1.00%)	N = 46 (3.27%)	0.111	0.47	0.29
II-6 L-4b	N = 418 (5.82%)	N = 73 (16.47%)	0.178	0.94	0.34
II-6 L-5 North	N = 828 (2.59%)	N = 50 (4.12%)	0.095	0.39	0.31
II-6 L-5 Center	N = 1,727 (5.42%)	N = 79 (6.52%)	0.227	0.75	0.38
II-6 L-5 South	N = 2,052 (6.44%)	N = 117 (9.66%)	0.335	1.06	0.44
II-6 L-6	N = 344 (2.80%)	N = 20 (7.96%)	0.120	0.42	0.36
II-6 L-7 (upper)	N = 705 (3.00%)	N = 55 (8.12%)	0.125	0.44	0.36
V-5	N = 1,233 (4.06%)	N = 53 (9.46%)	0.116	0.42	0.35
V-6 North	N = 31 (0.70%)	N = 4 (4.81%)	0.022	0.18	0.16
V-6 South	N = 217 (4.90%)	N = 8 (9.63%)	0.13	0.53	0.33

Percentages are the relative proportion of microartifacts within the high-density kernels, out of the total number of items of each category; length (L axis) depicts the longest axis of the kernel; width (W axis) stands for the longest axis perpendicular to L

Table 4.2 Counts of large burned flint items according to their proximity to high-density concentrations of burned flint microartifacts in different layers of GBY: C = concentration; FFT = flakes and flake-tools; CCT = cores and core-tools; PEB = pebbles

	Inside C			Close vicinity to C			Outside C			
Layer	FFT	CCT	PEB	FFT	CCT	PEB	FFT	CCT	PEB	Total
V-6	1	–	–	–	–	–	1	–	–	2
II-5	2	–	–	–	1	–	–	1	–	4
II-5/6	4	–	–	–	–	–	2	–	–	6
II-6 L-1	2	–	–	1	1	–	4	9	6	23
II-6 L-2	–	2	–	–	–	–	3	–	2	7
II-6 L-3	2	1	–	–	2	–	3	8	1	17
II-6 L-4	6	5	3	–	2	–	2	1	2	21
II-6 L-4b	1	–	–	–	–	–	1	3	4	9
II-6 L-5	1	–	1	1	1	–	3	6	2	15
II-6 L-6	2	1	2	1	1	–	4	–	1	12
Sub Total	21	9	6	3	8	–	23	28	18	116
V-5	–	–	–	–	–	–	-	1	–	1
II-6 L-7 (upper)	–	–	–	–	–	–	15	4	5	24
Total	21	9	6	3	8	–	38	33	23	141

the burned and unburned flints to be an indication for the origin (i.e., anthropogenic or natural) of the fire. *Model 2* considers a scenario in which the burned and unburned flint items are evenly distributed, so that we cannot rule out the possibility of a natural fire that deforms flint wherever it occurs. In *Model 3* the burned and unburned flints do not coincide and the burned items exhibit clusters. On the basis of various theoretical considerations (specified in Chapter 2), the significant differences in the distributions of burned and unburned flint items are considered firm evidence of an anthropogenic fire and the clusters of burned flint items are thus interpreted as remnants of hearths.

Several archaeological layers at GBY (e.g., II-6 L-1, II-6 L-2, II-6 L-3) display patterns of distribution similar to those suggested by *Model 3*, exhibiting evidence that unambiguously favors anthropogenic fire. Other archaeological levels (e.g., II-5, II-6 L-4b, II-6 L-6) demonstrate spatial patterning that better corresponds to *Model 2*, according to which we cannot rule out the possibility of a natural fire as the agent of the observed spatial patterns.

The following paragraphs will attempt a thorough inquiry into this possibility of natural fire. More specifically, we will inquire into the circumstances in which a natural fire would have ignited and spread through the Acheulian lake-shore occupations of GBY. In order to do so, we need to explore numerous issues related to fire ecology and fire dynamics. Insight into these issues is a prerequisite to visualizing the environmental conditions that would have allowed such a fire, as well as its effects on the archaeological material.

4.1 Fire Ecology

> We cannot assert that primitive man never saw a lightning fire anywhere. If he did so it was under most exceptional circumstances which did not encourage him to stay around and see what he could do with it. (Sauer 1961b:299)

Nearly half a century has passed since Sauer's thoughts on "Fire and Early Man" (Sauer 1961b), and we have learned that lightning strikes do not require "exceptional circumstances"; rather, they are common to such a degree that most fires throughout the world are started by lightning (Scott et al. 2000). This substantial development in our knowledge of fire dynamics is best illustrated by a debate conducted in the 1955 international symposium on "Man's Role in Changing the Face of the Earth" (Thomas 1956:410–422). The debate revolved around the significance of lightning as a triggering factor in the ignition of fire. Since lightning strikes had not been systematically documented, the most credible references were local newspapers reporting on local fires (ibid.:413). Thus, according to the available data documented at that time, "roughly a hundred thousand lightning strokes [occur] per day…" (ibid.:413), whereas the data available today document some eight million lightning strikes per day (Pyne et al. 1996; Scott et al. 2000).

Similarly, our knowledge of fire dynamics in different ecosystems is constantly growing and a large body of scientific data has accumulated through attempts to reconstruct the fire history of various ecosystems. These reconstructions use an assortment of methods, which include examination of historical records, analyses of tree-ring records, dating of fire-scar samples from standing trees, and analyses of charcoal within lake sediments (Pyne et al. 1996). Particularly essential is the use in paleoenvironmental reconstructions of microscopic charcoal, which incorporates fire histories and fire cycles of various chronological scales, as well as reflecting the impact of human manipulation of the environment through the use of fire. Examples of such studies can be found in the reconstruction of the fire history of Bolan Lake (Oregon) during the past 14,500 years using charcoal extracted from a sediment core (Briles et al. 2005), and in the 3,000-year-old charcoal record from Cliff Palace Pond (Kentucky) (Delcourt et al. 1998). There, a correlation between periods of human occupation and increase in local fires was observed, suggesting that the fires recorded by the charcoal are the result of human manipulation of the environment (Delcourt et al. 1998). Similarly, attempts to differentiate between natural fires and anthropogenic large-scale fires have been carried out mostly through charcoal analyses (e.g., Figueiral and Mosbrugger 2000; Haberle et al. 2001; Boyd 2002; Thevenon and Anselmetti 2007). Archaeological data can also contribute to our knowledge of fire history and suggest that humans have long conducted large-scale burning (e.g., for vegetation clearing or hunting purposes) in different areas of the world (Westbroek et al. 1993; see also a thorough review in Grayson 2001: Table V).

A variety of factors currently limits the interpretation of the fire history of the site of GBY. The diverse evidence of burned materials from the site (wood, charcoal, fruits, seeds, and flint) is adequate for the general assumption of a "fire history". However, the agent responsible cannot be inferred from the mere presence of these burned items. Nevertheless, the large body of scientific data that has accumulated, primarily through efforts to prevent *future* fire disasters, enables us to reconstruct a scenario of *past* natural fires at GBY. Adopting this line of thought, the following section attempts to evaluate the circumstances in which a natural fire may have ignited and spread at the site, thus contributing burned flint items to the archaeological layers. To do so, we should consider the various factors involved in the complex process of fire ignition, combustion, and behavior, available in the extensive literature on fire ecology.

4.1.1 Ignition and Combustion

Combustion is a rapid physical–chemical process in which the burning of plant material releases carbon dioxide, water, and the solar energy stored in the plant. These components are absorbed by plants through the process of photosynthesis, such that fire can be described as the reverse, rapid process of photosynthesis (Pyne et al. 1996:5; DeBano et al. 1998:20).

Although a wide range of fuels may be available for combustion, a common sequence of physical processes occurs in all fuels before the energy stored in them can be released and transferred. More specifically, three components are required for a fire to ignite and initiate combustion: *fuel*, *heat*, and *oxygen* (DeBano et al. 1998:20). Thus, combustible fuel must be available, enough heat must be applied to the fuel to raise its temperature to the ignition point, and there must be enough air to supply the oxygen needed to keep the combustion process going and maintain the heat supply for ignition of unburned fuels. These three components form the "fire triangle" that must be present or there will be no fire (DeBano et al. 1998:20).

Although spontaneous ignition[1] is possible, most fires require an ignition source. Lightning is the primary non-anthropogenic ignition source, resulting in forest, shrubland, and grassland fires (Pyne et al. 1996:56; DeBano et al. 1998:22). Although lightning is a major cause of wildfires, most lightning strikes do not in fact cause fires. Lewis (1989) differentiates between two types of lightning strikes in the Northern Rockies: "cold lightning", which accounts for about 80% of all cloud-to-ground discharges, and "hot lightning", which accounts for the remaining 20% and is the cause of most fires (Lewis 1989). In addition, during the 2 months in which electrical storms occur there, the ratio of strikes to fires can vary from as much as 1:1 to 1,000:1 or none, depending upon the object struck, the fuel conditions, and the weather (Lewis 1989:15). Similarly, Pyne et al. (1996) report on a 5-year study that demonstrated that only 0.01% to less than 0.001% of the cloud-to-ground lightning strikes actually started a wildfire (Fuquary 1962, cited by Pyne et al. 1996:19).

Lightning is also the major cause of wildfires in the Mediterranean zone (Whelan 1995), where the climate is a transitional regime between cold temperate and dry tropical climates. The Mediterranean Basin thus presents a unique climatic system of mild, dry summers and wet winters, with 90% or more of the annual precipitation falling within the 6 cool months (Naveh and Carmel 2004). The bimodal Mediterranean climate that we know today first appeared during the late Pliocene, at about 3.2 Ma, as part of a global cooling trend, and became firmly established throughout the region about 2.8 Ma (Blondel and Aronson 1999:21).

Similarly, floral and faunal indications have demonstrated that "the present-day climatic zones of Palestine, i.e. arid desert of the Jordan Valley in the East, Mediterranean seasonal climate of the hill country in the west, were … already developed in the Pleistocene" (Picard 1952:147). In addition, the Mediterranean wood species identified at GBY (Goren-Inbar et al. 2002b), and other paleobiological evidence (e.g., remains of mollusks, crabs, fish, and mammals), strongly suggest that the seasonal climate pattern at the time of deposition resembled the pattern seen in the present-day Hula Valley.

[1] Spontaneous ignition is likely to evolve in large (greater than 1 m) piles of thin fuels (i.e., hay, chip, and sawdust) with a moisture content greater than 20% and soil mixed into the pile (Pyne et al. 1996:20).

The most probable type of natural fire in the Mediterranean region is surface wildfire (Whelan 1995), resulting from natural ignition and combustion. Even when natural ignition such as lightning does occur, it requires appropriate climatic conditions in order to combust and evolve into a wildfire. In the present-day Hula Valley, lightning storms are most common from October to March (data from the Israel Meteorological Service). However, lightning occurring at that time of year (the rainy season) will rarely produce spontaneous fires (Whelan 1995:26).

In our attempts to reconstruct the hypothetical circumstances of a natural fire at GBY, we can thus suggest lightning as a possible ignition source. However, the probability of ignition varies among different environments. Probability of ignition is the chance that an ignition will result if a firebrand lands on flammable material and is defined as a function of fuel moisture and fuel temperature (Pyne et al. 1996:56). These definitions are based on experiments in which lighted matches were dropped onto fuel beds with various levels of moisture content (Blackmarr 1972, cited by Pyne et al. 1996:56). Certainly, the energy supplied by lighted matches cannot compete with the energy discharged by lightning strikes. In addition, these experiments considered a particular type of fuel (slash pine litter). However, they do demonstrate the fact that it takes more than an ignition source to set a fire and that, following the ignition point, various environmental factors (e.g., fuel moisture) are involved in determining whether heat will transfer and fire will occur. These are discussed in the following paragraph.

4.1.2 Heat Transfer

The mere presence of a heat source does not necessarily result in a fire; heat must be transferred from the fire source to the unburned fuel for a fire to continue to burn. Although heat is transferred in all directions, large amounts of heat are lost into the atmosphere and a significantly smaller amount of the heat is transferred to unburned fuels to sustain the fire (Pyne et al. 1996:12; DeBano et al. 1998:31). Thus, even if ignition has occurred, fire encounters a second obstacle in the efficient transfer of the heat from the ignition source to nearby fuels.

The efficiency of combustion is largely determined by the moisture level of the fuel and its surrounding soil. At GBY, all occupation episodes are associated with lake-margin deposits and are waterlogged. The permanently moist conditions of the GBY deposits have enabled the preservation of botanical remains throughout the entire stratigraphic sequence. This is extremely significant, since sediment moisture appears to have a substantial influence over the dynamics of heat transfer. In wet deposits, like those of the waterlogged site of

GBY, moisture reduces underground temperatures. When the surface temperature of the fire exceeds the boiling point of water, evaporation delays heating of the underlying soil (Whelan 1995). Consequently, moisture increases the amount of heat required. Before a fire can become self-sustaining, sufficient heat must be absorbed by fuels to evaporate much of the water and make them flammable. Accordingly, "excess heat-absorbing water … can result in a failure of a fire to ignite" (DeBano et al. 1998:21). Moisture in fuels increases their ignition time and decreases their burning rate. In short, "dry fuels burn hot, completely and quickly, while moist fuels either do not burn or do so slowly and at lower temperatures" (DeBano et al. 1998:28).

Waterlogged environments (e.g. peatlands, bogs, and marshes) can be subject to fires. However, fire in wetlands and riparian ecosystems occurs following a period of drought that reduces the water table and allows burning of the available fuel (DeBano et al. 1998:229–243).

Let us assume a scenario at GBY in which, despite the waterlogged sediments, heat was efficiently transferred and moist fuels evaporated water to a degree that allowed them to combust and burn. Will our fire have produced sufficiently high temperatures to burn the flint artifacts scattered on the surface? In order to address this question, we should consider aspects of fire behavior that determine the impact of fire on various environmental settings.

4.1.3 Fire Behavior

This section discusses the final stage of our hypothetical reconstruction of a natural wildfire at GBY. This fire began with a lightning strike, which has managed to overcome the moisture of the sediments and produce an ignition source. Subsequently, the ignition source has been able to transfer sufficient heat to the surrounding sediments, initiating combustion. We should now consider the possible intensity of such a fire, taking into account the environmental setting of GBY.

The intensity of a fire varies vertically (i.e., different temperature profiles above and below ground). The vertical distribution of temperatures in a fire is determined by several factors, including distribution of the fuel, wind speed, and direction of the fire front (i.e., head fire or back fire) (Whelan 1995). Although we clearly cannot determine these factors from the archaeological record, general data regarding measurements of peak temperatures above and below ground are of great importance when attempting to evaluate the origin of a fire.

Most available measurements record only peak temperatures and not duration of temperature, and indicate that above-ground temperature declines with height (Whelan 1995:15). Whelan further reports on a study that measured peak temperatures of a fire in several locations and at three different heights (Trollope 1984, cited by Whelan 1995:15). According to that study, the highest temperatures in both head and back fires occurred in the grass canopy. This and other studies suggest that in fires in grassland and shrubland, peak temperatures reach 200–300°C (light to moderate fires) and even 500–600°C or more (heavy fires) (Whelan 1995: Fig. 2.3).

Based on various experiments conducted on the burning of flint (discussed in Section 2.3.2), these recorded temperatures would have been sufficient to burn flint items at GBY. Yet, others have suggested that little damage to lithic artifacts is likely to occur during wildfires of low severity and no adverse effects are likely to occur below 500°C (DeBano et al. 1998). In addition, "… most of the damage to these artifacts occurs where there are surface accumulations of fuels, such as downed logs, that are burned. The resulting high and long-duration temperatures can cause sooting, potlids, and some crazing of the artifacts" (DeBano et al. 1998:271).

The intensity of a fire and thus its eventual effects are influenced by a variety of parameters, including the amount of available fuel, moisture level of the fuel and sediments, local weather conditions, and the topography of the burning site (Whelan 1995; Pyne et al. 1996; DeBano et al. 1998). The following section attempts to examine the ecological data on fire ignition, combustion, and behavior in relation to the archaeological site of GBY.

4.1.4 Probability of a Natural Fire at GBY

This discussion attempts to reconstruct the possible history of the fire regime in the area of GBY ca. 790,000 years ago. Common methods such as the analyses of charcoal records, tree-rings, or fire-scar samples are not available to us. However, the above review has provided valuable data on the required circumstances in which a natural fire would have ignited and spread at GBY (see also Appendix 4).

The data on the various environmental parameters involved in a natural fire reject the scenario of a natural fire at GBY. The Mediterranean climatic system and the waterlogged deposits of the site do not conform with the conditions required by a natural fire to ignite and spread. Thus, the combination of the three components of the "fire triangle" (*oxygen*, *fuel*, and *heat*) was most probably absent from the lake-shore occupations of GBY. Apart from the air (*oxygen*) required to keep the combustion process going, fire needs large quantities of combustible *fuel*. At GBY abundant fuel (i.e., the wood assemblages) was recovered from the excavated archaeological occurrences but was rarely found burned; according to Goren-Inbar et al. (2002b), less than

1% of the wood segments from GBY (14 out of 1,568) are burned. Finally, considering the constantly damp conditions at the site, a natural ignition source (e.g., lightning) most likely could not apply enough *heat* to the fuel to raise its temperature to the ignition point.

In summary, the necessary conditions for a natural fire at GBY would have had to include the following (see also Appendix 4):

1. An ignition source (most likely a lightning strike).
2. A dry interval in the cold and wet season (during which lightning strikes occur in the Mediterranean zone).
3. Abundant availability of combustible fuel to allow heat transfer.
4. Extremely high combustion temperatures to allow evaporation of the soil's moisture.

In the event that such a fire indeed occurred, we would expect this fire to:

1. Consume the available fuels on the surface.
2. Practically "boil" the sediments through evaporation of the soil's moisture.
3. Extensively damage the flint artifacts scattered on the surface in direct contact with the fire, while exceeding the temperatures required to alter flint (ca. 350ºC). These would be particularly large items (e.g., macroartifacts and pebbles), which are exposed on the surface, rather than smaller ones (e.g., microartifacts), which are slightly submerged in the wet surface.
4. Finally, considering the fact that fire occurs sequentially throughout the stratigraphic sequence at GBY, we are obliged to consider this scenario as characteristic of a fire regime in which such exceptional circumstances occurred repeatedly, frequently setting off natural fires on the lake margin.

The archaeological data, however, are not in accordance with such a scenario, whether of a natural fire that ignited and spread during the time of occupation (i.e., surface fire with all archaeological material scattered on the surface) or of a fire that ignited and spread after deposition (i.e., subsurface fire with all archaeological material buried underground at various depths).

Surface Fire: If we assume that a wildfire occurred during (or immediately following) occupation, we must take into consideration the fact that the highest temperatures in such fires can reach 550°C (Whelan 1995), hot enough to damage organic material as well as flint items. If surface wildfires were responsible for the burning of the organic and inorganic material at GBY, we would expect to find high frequencies of burned material. The GBY layers yielded large quantities of unburned wood interpreted as driftwood (Goren-Inbar et al. 2002b), an excellent fuel that would have fanned any wildfire, thus increasing the frequencies of burned material. However, less than 1% of the wood segments (Goren-Inbar et al. 2002b) and less than 2% of the carbonized wood pieces (Goren-Inbar et al. 2004) are burned. A low percentage of burning is also recorded in the different lithic categories of the excavated flint pieces (see Chapter 3). Flint pebbles, which in most cases do not show any signs of burning (a range of 0.00–1.94% of the flint pebbles within all layers), exhibit particularly low frequencies of burning in comparison to the flint macroartifacts (0.31–6.06%) and microartifacts (0.76–5.80%). It is interesting to note that comparable frequencies of burning (i.e., 4% of the microartifacts [smaller than 1 cm]) are recorded at Upper Paleolithic sites in Western Europe, where the hearths are particularly visible structures (Leesch et al. 2005; Plumettaz 2006). In addition, the frequencies of burning among the large flint items, as well as their spatial configuration, further undermine a scenario of natural fire. With the exception of Layer II-5/6, all the GBY layers exhibit higher relative frequencies of burning in the CCT category (i.e., cores and core-tools) in comparison with the FFT category (i.e., flakes and flake-tools) (see tables in Chapter 3). Yet, when examining the spatial association between those burned items and the concentrations of burned flint microartifacts, it seems that burned FFTs tend to be found within those clusters more often than CCTs and pebbles (Table 4.2).

If surface wildfires were responsible for the burning of the flint items, we would expect to find higher frequencies of burning amongst larger items, which lie on the surface, than amongst smaller, thinner, items, which tend to be submerged in the sediment. Indeed, the CCT category exhibits higher frequencies of burning. However, flint pebbles, which like the CCTs are thick lumps of rock and should display higher frequencies of burning, exhibit particularly low frequencies of burning. A natural wildfire would not differentiate between pebbles and CCTs while burning larger lumps of flint items on the surface. The variability of burning frequencies observed amongst the different lithic categories, and the spatial distribution of the large burned flint items, are more likely the result of human activities than of natural wildfires. While drawing on the above assertions, however, we should take into account the fact that the large burned flint items generally occur in relatively small numbers and do not provide sufficient sample sizes for in-depth analyses.

Subsurface Fire: The scenario of burning of the archaeological material subsequent to deposition is a possible one for areas where there are heavy accumulations of organic matter that can undergo a ground fire, burning deeply into the organic material above and within the soil profile (DeBano et al. 1998:56). Various studies have indicated that peak temperatures at 2.5 cm below the surface are likely to be well below 100°C, even when the fire above is of very high intensity (Whelan 1995:17). Drawing from experimental work on the effects of fire on flint (see Section 2.3.2), it is clear that the burning of the flint items at GBY required on the one hand an intensive fire with relatively high temperatures, and on the other hand direct contact with the flints. The facts that fire damaged relatively low frequencies of flint items and that the burned flint microartifacts are found spatially clustered are suggestive of an anthropogenic rather than a natural fire. More specifically, the low underground temperatures described above are unlikely to have damaged subsurface flint items at the site.

Based on these considerations, the combination of the ecological data and the archaeological record allows the firm rejection of the possibility of recurrent natural fires at the Acheulian lake-shore occupations of GBY.

4.2 Use of Fire at GBY

Burned flints are recorded within each of the archaeological horizons at GBY. In all of the analyzed assemblages, the frequencies of burned artifacts are low and the burned flint microartifacts exhibit various patterns of clustering and are unevenly distributed throughout the occupation surfaces. Examination of the distribution patterns of the burned and unburned flint items in the various archaeological layers (see Chapter 3) has demonstrated that the main differentiating parameter is the degree of overlap between the burned and unburned flint microartifacts. Thus, while in some layers the burned and unburned flint microartifacts overlap each other to different extents (either moderately or entirely), in others they exhibit clustering at distinctly different locations.

When constructing the hypotheses of this study, we chose to be extremely cautious in analyzing the distribution patterns of the burned flints. Such circumspection was essential for the careful distinction between patterns of anthropogenic and natural fires, particularly when attempting to identify early hominin use of fire. The three possible models were discussed earlier in this chapter. *Model 1*, which postulates an absence of burned flint items, has been confidently rejected. In cases in which a significant overlapping between the distribution of the burned and unburned flints is observed, it has been determined that we cannot rule out the possibility of a natural fire that damaged flint wherever it occurred (i.e., *Model 2*). Thus, only cases in which the burned and unburned flints do not coincide and the burned flint microartifacts are clustered in distinctively different areas have been considered to represent evidence of an anthropogenic fire (i.e., *Model 3*).

The fire ecology data (reviewed above) have established a solid basis for rejection of the hypothesis of natural fire during the Acheulian occupations at GBY. In addition, it is clear at this point that several layers within the GBY occupational sequence (e.g., II-6 L-1, II-6 L-2, II-6 L-3) exhibit distribution patterns that accord with *Model 3* and are thus unambiguously in favor of anthropogenic fire. These archaeological occurrences indicate that the knowledge and technology of fire use were at hand during the Acheulian occupations at GBY.

We should, however, inquire into the cases in which the patterns of distribution of the burned and unburned flints are not consistent with *Model 3*. These are layers that exhibit overlapping distribution patterns of the burned and unburned flint microartifacts. This is the main objective of the following discussion.

4.2.1 Differentiation in the Distribution of Burned and Unburned Flint

A fundamental assumption in the framework of this study is that the aggregation of human activities in vicinity to hearths results in denser concentrations of small waste products in situ, and thus that clustering of burned waste is evidence for the location of ancient hearths (see Section 2.2). Where hominins carried out flint knapping activities in several locations on the occupational surface, with only some of these in proximity to hearths, the distribution of the burned and unburned flint will not entirely coincide. However, where knapping activities were confined to the hearth(s) area, we will expect to find overlapping of the burned and unburned flint items. Such circumstances are not exceptional at sites in which hearths were used as a focal point of activities. An example of this can be found at the Magdalenean sites of Champréveyres and Monruz, where Leesch et al. (2005) have repeatedly observed that "...the burned and unburned flint chips ... are found regularly together in the hearth residues" (Leesch et al. 2005:6). A detailed comparison between the frequencies of burned and unburned flint microartifacts within the hearths further suggests that the frequencies of unburned flint are often higher than those of the burned ones (Plumettaz 2007: Fig. 3).

The following section examines the characteristics of the various occupational layers in relation to the degree of observed spatial differentiation between the distributions of the burned and unburned flint microartifacts (Table 4.3).

4.2.1.1 Highly Significant Differentiation

Three archaeological layers (II-6 L-1, II-6 L-2, II-6 L-3) exhibit a spatial clustering of the burned flint microartifacts that is significantly different from that of the unburned ones, as well as other parameters (e.g., statistical tests; random plotting tests; see Chapter 3 for a thorough presentation)

Table 4.3 Attributes of the GBY layers relative to the degree of spatial differentiation between burned and unburned flint microartifacts; dimensions are in m (LEN and WID axis) and m^2 (area)

	Entire layer								Kernel of cluster					[c]SSQ Encircling Kernel	
	Frequencies of burned flint			[a]BFM percentage per sub-square			Chi square test of BFM		Dimensions		[b]Percentage of BFM and UBFM				
	Microartifacts N (%)	Macroartifacts N (%)	Pebbles N (%)	Mean; maximum	Std	[d]Relative area BFM	Σx^2	*df*; *p*	LEN axis; WID axis	Area	% BFM	% UBFM	Ratio % BFM/UBFM	OBS – EXP % BFM	SR for BFM; n (expected)
Highly significant differentiation															
II-6 L-1	754 (1.40)	17 (1.40)	6 (0.25)	0.34; 5.03	0.84	0.63	580.57	$df = 90$; $p < 0.001$							
N									0.46 0.41	0.148	3.97	1.25	3.17	2.83	5.25; n=16.6
S									0.46 0.31	0.117	3.71	0.59	6.28	3.32	7.60; n=10.9
II-6 L-2	563 (0.76)	5 (1.05)	2 (0.25)	0.31; 16.30	1.28	0.42	913.27	$df = 69$; $p < 0.001$	0.51 0.40	0.165	9.41	2.39	3.39	12.63	15.79; n=20.5
II-6 L-3	877 (0.95)	16 (3.13)	1 (0.09)	0.69; 10.30	1.48	0.55	673.37	$df = 66$; $p < 0.001$	0.50 0.41	0.161	6.72	2.13	3.15	6.61	10.18; n=32.2
Relative overlapping															
V-6	83 (1.84)	2 (0.70)	00 (0.00)	2.00; 20.50	3.63	0.48	56.51	$df = 31$; $p < 0.01$							
N									0.18 0.16	0.022	4.81	0.70	6.87	1.10	0.40; n=5.08
S									0.53 0.33	0.13	9.63	4.90	1.96	10.00	2.80; n=8.7
II-5/6	400 (4.50)	6 (5.30)	00 (0.00)	0.39; 8.25	1.23	0.39	204.48	$df = 64$; $p < 0.001$							
N									0.97 0.23	0.167	8.00	4.59	1.74	−2.25	-1.40; n=42.1
S									0.40 0.38	0.122	4.50	1.99	2.26	2.26	2.08; n=18.9
II-6 L-4	1,406 (1.32)	16 (2.45)	5 (0.85)	0.59; 6.82	1.25	0.60	1063.8	$df = 60$; $p < 0.001$							
N									0.41 0.37	0.121	3.34	1.79	1.86	2.38	4.71; n=50.5
SE									0.47 0.36	0.137	5.83	1.67	3.49	2.64	22.48; n=2.7
SW									0.47 0.29	0.111	3.27	1.00	3.27	1.58	3.90; n=31.7
II-6 L-7 (upper)	677 (2.80)	19 (3.07)	5 (0.63)	0.51; 9.89	1.34	0.77	109.32	$df = 57$; $p < 0.001$	0.44 0.36	0.125	8.12	3.00	2.70	4.27	4.70; n=38

	Entire layer								Kernel of cluster					[c]SSQ Encircling Kernel	
	Frequencies of burned flint			[a]BFM percentage per sub-square			Chi square test of BFM		Dimensions		[b]Percentage of BFM and UBFM				
	Microartifacts N (%)	Macroartifacts N (%)	Pebbles N (%)	Mean; maximum	Std	[d]Relative area BFM	Σx^2	*df*; *p*	LEN axis; WID axis	Area	% BFM	% UBFM	Ratio % BFM/UBFM	OBS – EXP % BFM	SR for BFM; n (expected)
Almost complete overlapping															
V-5	560 (1.81)	1 (0.31)	00 (0.00)	1.81; 13.40	2.87	0.80	176.43	$df = 33; p < 0.001$	0.42 0.35	0.116	9.46	4.06	2.33	0.79	0.55; n=70.3
II-5	162 (4.48)	4 (3.41)	00 (0.00)	0.34; 11.10	1.45	0.21	200.43	$df = 73; p < 0.001$	0.61 0.51	0.219	10.5	7.73	1.35	−1.19	-0.44; n=20
II-6 L-4b	443 (5.80)	5 (3.20)	4 (1.94)	0.69; 14.40	1.95	0.49	386.27	$df = 55; p < 0.001$	0.94 0.34	0.178	16.5	5.82	2.82	4.30	2.86; n=44.8
II-6 L-5	1,211 (3.66)	12 (6.06)	3 (0.78)	0.69; 6.77	1.38	0.76	241.03	$df = 54; p < 0.001$							
N									0.39 0.31	0.095	4.12	2.59	1.59	−0.24	-0.31; n=84.9
C									0.75 0.38	0.227	6.52	5.42	1.20	−0.84	-1.16; n=77.2
S									1.06 0.44	0.335	9.66	6.44	1.50	−0.09	-0.16; n=46.1
II-6 L-6	251 (2.00)	9 (2.69)	3 (0.64)	0.69; 7.96	1.58	0.60	288.06	$df = 57; p < 0.001$	0.42 0.36	0.120	7.96	2.80	2.84	2.53	1.73; n=13.6

BFM = burned flint microartifacts; UBFM = unburned flint microartifacts; SSQ = sub-square; N = north; C = center; S = south; E = east; W = west; LEN = length; WID = width; OBS = observed; EXP = expected; SR = standardized residual

[a]Statistics for percentage of burned flint microartifacts per excavated unit/sub-square (0.5 m^2)

[b]Percentages are the relative proportion of microartifacts within the high-density kernel of the cluster, out of the total number of microartifacts in each category (burned or unburned)

[c]The sub-square that either encircles or occupies the majority of the high-density kernel of the cluster

[d]The ratio between the number of excavated units in which burned flint microartifacts occur and the total number of excavated units of the layer

that support the observed spatial differentiation and are thus suggestive of an anthropogenic fire. The characteristics of these layers include:

1. A low percentage of burning amongst the flint items. The percentage of burned microartifacts in these layers is in the range of 0.76–1.40%, burning of flint macroartifacts is in the range of 1.05–3.13%, and flint pebbles exhibit particularly low percentages of burning (0.09–0.25%) (Table 4.3).
2. The spatial distribution of burned flint macroartifacts and pebbles demonstrates that in these layers most of the burned flint macroartifacts, and all the burned flint pebbles, occur outside the high-density concentrations of burned flint microartifacts (Figs. 3.28, 3.33, 3.38; Table 4.2). Thus, in these layers several large burned items occur within and in close proximity to the high-density concentrations but are not confined to these areas; rather, they seem to be scattered throughout the occupation surfaces.
3. The burned flint microartifacts do not occur throughout the entire excavated surface but occupy a smaller area. The ratio between the number of excavated sub-squares in which burned flint microartifacts are recorded and the total number of excavated sub-squares ranges in these layers between 0.42 and 0.63 (Table 4.3).
4. Within these layers, examination of the percentages of burned flint microartifacts in each of the excavated units (Figs. 3.24, 3.29, 3.34) demonstrates that the spatial distribution of the burned flint microartifacts is uneven, and that several sub-squares exhibit higher frequencies. Accordingly, a significant deviation is observed in these layers between the mean percentage (with a range of 0.31–0.69%) and the maximum percentage (with a range of 5.03–16.30%) of burned flint microartifacts per excavated unit (Table 4.3).
5. The kernel density maps demonstrate that in these layers the burned flint microartifacts are clustered. A single cluster is recorded in Layer II-6 L-2 and in Layer II-6 L-3, while in Layer II-6 L-1 two adjacent clusters are recorded (Figs. 3.25, 3.30, 3.35).
6. The distribution patterns of the burned flint microartifacts in these layers are notably different from those of the unburned ones. In addition, since the latter occur in much higher frequencies, the random density maps exhibit greater similarity to the distribution patterns of the unburned flint microartifacts than to those of the burned ones (Figs. 3.26, 3.31, 3.36).
7. A chi square test on the burned flint microartifacts of these layers suggests that their distribution is significantly different from an expected, uniform distribution (i.e., $p < 0.001$) (Table 4.3).
8. Calculation of Standardized Residuals (SR) on the burned flint microartifacts of these layers demonstrates that significant SR values are recorded within the burned clusters (Table 4.3). This suggests that the concentrations of burned flint microartifacts in these areas are the major contributors to the observed spatial differentiation.
9. In the sub-squares that encircle the high-density kernels of the burned clusters, the observed percentage of burned flint microartifacts is higher than the expected percentage. The deviation between the observed and expected percentage of burning in these areas ranges between 2.83 and 12.63 (Table 4.3).
10. Within the area of the high-density kernel of the clusters, the percentage of burned flint microartifacts is higher than that of the unburned ones. The ratio between the percentage of the burned flint microartifacts and the percentage of the unburned ones within the high-density kernel of the clusters ranges in these layers between 3.17 and 6.28 (Table 4.3).

4.2.1.2 Relative and Complete Overlapping

For most of the analyzed archaeological layers, the burned and unburned flint microartifacts coincide with each other to various extents. In some layers (Layers V-6, II-5/6, II-6 L-4, II-6 L-7) the patterns of distribution overlap only partially, while in others (Layers V-5, II-5, II-6 L-4b, II-6 L-5, II-6 L-6) they overlap almost completely. However, when comparing the different parameters that characterize the "highly significant differentiation" patterns (listed above) to those of the "relative and complete overlapping" patterns, the similarities appear to be larger than the discrepancies (Table 4.3). Thus, the percentage of burning amongst the flint items is relatively low, particularly within the flint pebbles, which rarely exhibit burning damage, and when they do it is in very low frequencies (Table 4.3). The burned flint microartifacts are never evenly distributed throughout the exposed surface, resulting in a considerable deviation between the mean and maximum percentage of burning per excavated unit (Table 4.3). In most cases, the relative area in which the burned flint microartifacts occur is no higher than that recorded for the "highly differentiated" patterns (three exceptions are Layers V-5, II-6 L-5, II-6 L-7). Apart from one layer (Layer V-6, which has the smallest sample of burned flint microartifacts: N = 83), the chi square test on the burned flint microartifacts of these layers suggests that their distribution is significantly different from an expected, uniform distribution (Table 4.3). Furthermore, within the high-density kernel of the burned clusters, the relative percentage of burned flint microartifacts is always higher than that of the unburned ones (Table 4.3).

Other parameters, which derive from the correlation between the distribution of the burned and the unburned flint microartifacts (e.g., observed vs. expected values of burning,

SR values, random distribution patterns), are less significant in these layers (Table 4.3). A particularly prominent example of this can be found in the patterns exhibited in Layer II-6 L-5. There, the burned flint microartifacts occur throughout most of the excavated surface. The density maps of the burned and unburned flint microartifacts overlap almost completely, and both exhibit three dense areas with only minor differences observed within the configuration of the clusters (Fig. 3.50). Accordingly, the patterns illustrated by the random density maps are similar to those of the observed distribution of the burned flint microartifacts (Fig. 3.51). The chi square test of the burned flint microartifacts of Layer II-6 L-5 suggested that their observed distribution is significantly different from an expected, uniform distribution. However, calculations of standardized residuals for different excavated units imply that the major contributors to this are sub-squares that lie outside the concentrations of burned items. Despite the significance of the SR values, these sub-squares exhibit relatively minor deviations between the observed and expected percentages of burned flint microartifacts (Fig. 3.52). Nonetheless, amongst the three dense clusters of burning in Layer II-6 L-5, the high-density kernel exhibits higher percentages of burned flint microartifacts than those of the unburned ones (Table 4.3).

4.2.1.3 Levels of Differentiation: Summary

Burned flint items occur in all the Acheulian occupations at GBY. Analyses of their spatial distribution have demonstrated that these are not evenly distributed throughout the excavated surfaces, and dense clusters of burned flint microartifacts were repeatedly identified. These clusters are relatively small and incorporate both burned and unburned flint microartifacts. However, the relative percentage of the burned flint microartifacts within the high-density kernels of these clusters is always higher than that of the unburned ones.

When comparing the distribution patterns of the burned and unburned flint microartifacts in each of the assemblages, the patterns were found to exhibit different degrees of overlapping, from complete overlap to distinct differentiation. This spatial variation was initially used to distinguish between layers in which the evidence of hominin use of fire is highly significant (i.e., the distinct differentiation of *Model 3*) and those where the evidence may equally be the result of anthropogenic or natural fires (i.e., the overlapping patterns of *Model 2*).

A variety of ecological considerations, specified above, reject a scenario of natural fires as the agent responsible for the burning of the flint items in the different occupational horizons at GBY. Thus, the joint occurrence of two different spatial patterns (i.e., overlapping vs. differentiation of burned and unburned flints) is interpreted here as reflecting the variability of the spatial patterning of activities throughout the occupational horizons. Specifically, the results of this study suggest that in some cases flint knapping was carried exclusively in vicinity to hearth(s), so that the distributions of the burned and unburned flint microartifacts coincide, and in other cases flint knapping was not confined to the hearth area, so that the burned and unburned flint microartifacts are distributed in distinctively different clusters.

The spatial complexity of activities amongst the GBY layers, which is demonstrated in the distribution of the burned and unburned flint microartifacts, is further illustrated by the spatial configuration of other components of the archaeological material. A thorough spatial analysis of these other components, which will surely provide valuable data on the behavioral complexity of the Acheulian hominins of GBY, is beyond the scope of this study. The role that fire may have played within the various activities is discussed in the following section.

4.3 Patterns of Human Activities and Fire at GBY

The different archaeological layers at GBY are all components of the same cultural complex and share similar cultural characteristics. However, despite the general similarities, the various archaeological layers exhibit different patterns of activity. The remnants of these activities can be found primarily in the lithic assemblages (e.g., emphasis on the production of particular artifacts) and the faunal assemblages (e.g., butchering patterns).

The previous chapters have established evidence indicative of the use of fire by the hominins of GBY. In the following discussion we will investigate the role played by fire in relation to different activities carried out within different occupational levels of the site. More specifically, we will examine whether the use of fire is associated with particular tasks or is an integral part of the cultural repertoire, regardless of the type of activity. In order to achieve this goal, several archaeological layers at GBY for which data are published and available for such a discussion are presented.

4.3.1 The Carcass-Processing Site of Layer V-6

The excavation of Layer V-6 (Fig. 4.1) yielded extremely rich, dense, and remarkably well-preserved faunal assemblages (Rabinovich et al. 2008, in press), together with a flint-dominated lithic assemblage and paleobotanical remains of wood, bark, fruits, and seeds.

Fig. 4.1 General view of the excavations of Layer V-6

Several considerations indicate that the excavated surface represents the original spatial configuration of the archaeological material. The depositional setting of the archaeological material is at the interface between fine-grained offshore muds and a mollusk coquina. This setting is the outcome of hominin activities that were carried out on a land surface following a drop in the lake level; the occupational surface was buried shortly afterwards due to a rise in the lake level (Feibel 2001). This sealing, together with negligible taphonomic disturbances, left the archaeological material in an excellent state of preservation: the flint artifacts are in mint condition, no significant signs of weathering are present on the bones, the assemblages of fossil freshwater crabs include complete mandibles and pincers in their original anatomical configuration (Ashkenazi et al. 2005), and several bones have been refitted.

Unlike the cultural sequence represented in Area B (e.g., Layer II-6 L-1–7), the lithic assemblage of Layer V-6 (as well as that of Layer V-5) is characterized by a predominance of flint artifacts and a paucity of bifacial tools. A detailed techno-typological analysis of the lithic assemblage identified several typical products of handaxe manufacture, demonstrating that flint handaxes were manufactured during the time of occupation of Layer V-6, introduced as finished tools, and exported from the site (Goren-Inbar and Sharon 2006).

An interesting pattern emerges from the spatial distribution of the flint-dominated lithic assemblage. The percentage of flint is significantly higher than that of basalt or limestone amongst microartifacts (N = 6,585, of which 68.44% are flint, 30.09% basalt, and 1.45% limestone) and amongst FFTs (N = 322, of which 85.40% are flint, 14.28% basalt, and 0.31% limestone). Limestone artifacts are not recorded, and basalt is more dominant than flint within the lithic categories of CCTs (N = 28, of which 32.14% are flint and 67.85% basalt) and bifacial tools (N = 6: a single flint handaxe and five basalt cleavers) (see Section 3.2, Table 3.2).

The dominance of basalt within the bifacial tools category is interesting, particularly in view of the abundance of flint flakes that are typical of the production of handaxes (discussed above and see Goren-Inbar and Sharon 2006). In this context, it is worth examining the spatial distribution of the basalt and flint microartifacts and macroartifacts. Figure 4.2 shows that the basalt microartifacts of Layer V-6 (N = 1,982) occur in two adjacent concentrations in the northern part of the excavated surface. These concentrations do not coincide with the concentrations of burned and unburned flint microartifacts (Fig. 3.7). Furthermore, examination of the spatial distribution of basalt macroartifacts in comparison with the density map of basalt microartifacts suggests that most macroartifacts, particularly the cleavers, are spatially associated with the concentrations of basalt microartifacts (Fig. 4.2). This pattern differs from that displayed by the distribution of the flint artifacts, in which both microartifacts and macroartifacts seem to be scattered throughout the exposed surface (Fig. 4.2).

Several conclusions can be drawn from the spatial patterning of the lithic assemblage:

1. The concentrations of basalt microartifacts in the northern part of the excavated area demonstrate an emphasis on basalt knapping in that area.
2. This knapping likely involved the manufacturing of basalt cleavers (only two of which are represented in the field map; Fig. 4.3) as well as basalt flakes, flake-tools, and core-tools.
3. The manufacturing of flint artifacts does not seem to be restricted to a particular area and the products of flint knapping occur throughout the exposed surface.
4. The location of the high-density concentrations of burned flint microartifacts in the central part of the excavated surface cannot be spatially associated with the knapping of basalt or that of flint. The northernmost concentration

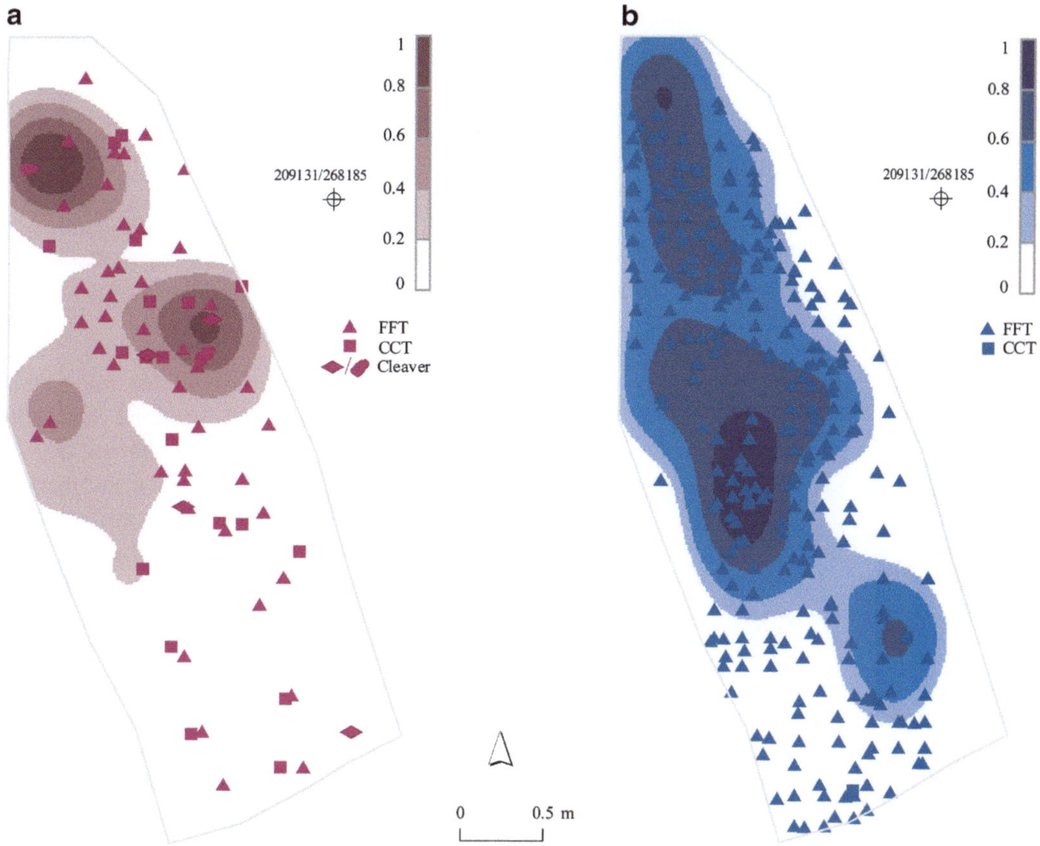

Fig. 4.2 Layer V-6. (**a**) Kernel density map of basalt microartifacts and the distribution of basalt macroartifacts and (**b**) kernel density map of unburned flint microartifacts and the distribution of flint macroartifacts

of basalt microartifacts seems to be more isolated from the central area, where the use of fire is recorded. The second, southern, high-density concentration of basalt microartifacts, which is spatially associated with three basalt cleavers, is, however, in close vicinity to a high-density concentration of burned flint microartifacts.

The wood assemblage of Layer V-6 comprises 26 fragments, on which no signs of burning were observed. Taxonomic identification was carried out on 21 fragments, which include ash (*Fraxinus syriaca*), willow (*Salix* sp.), cedar (*Cedrus* sp.), oak (*Quercus* sp.), and juniper (*Juniperus* sp.) (Goren-Inbar et al. 2002b: Tables 4 and 9). Of the 26 wood fragments, 23 are drawn in the field map (Fig. 4.3). The spatial distribution of these unburned wood fragments demonstrates that they occur throughout the entire exposed surface, including the area within and in the vicinity of the high-density concentrations of burned flint microartifacts (Fig. 4.3).

The diverse species represented in the faunal assemblage of Layer V-6 consist primarily of fallow deer (*Dama*), as well as horse, hippopotamus, and elephant. The excellent state of preservation of the faunal assemblage permitted the performance of detailed taxonomic, morphometric, taphonomic, and microscopic analyses (Rabinovich et al. 2008, in press).

Traces of animal-induced modifications (i.e., by rodents and carnivores) are rare and are discerned on only 2.8% of the entire bone assemblage of Layer V-6 (NISP = 1,248). Detailed analyses of hominin-induced damage were carried out on the skeletal elements of *Dama* (NISP = 504; MNI = 4) and on bones assigned to a *Dama*-sized species (NISP = 293). A large variety of modifications was observed within the *Dama* and *Dama*-sized groups and included cut marks (6.6%; NISP = 53) and percussion marks (9.6%; NISP = 77); the latter are conical fractures and bone flakes that constitute evidence of marrow extraction (Rabinovich et al. 2008).

These analyses indicated that systematic butchering of *Dama* was practiced by the Acheulian hominins of GBY. It involved skinning (evidenced by cut marks on crania, metapodials, and phalanges), disarticulation (evidenced by cut marks and hack marks on first cervical vertebrae, mandibles, and limb elements), defleshing and filleting (evidenced by cut marks on mandibles, vertebrae, ribs, scapulae, humeri, radii, ulnae, pelves, femora, and tibiae), and marrow extraction (evidenced by percussion marks on humeri, radii, metapodials, femora, tibiae, and phalanges)

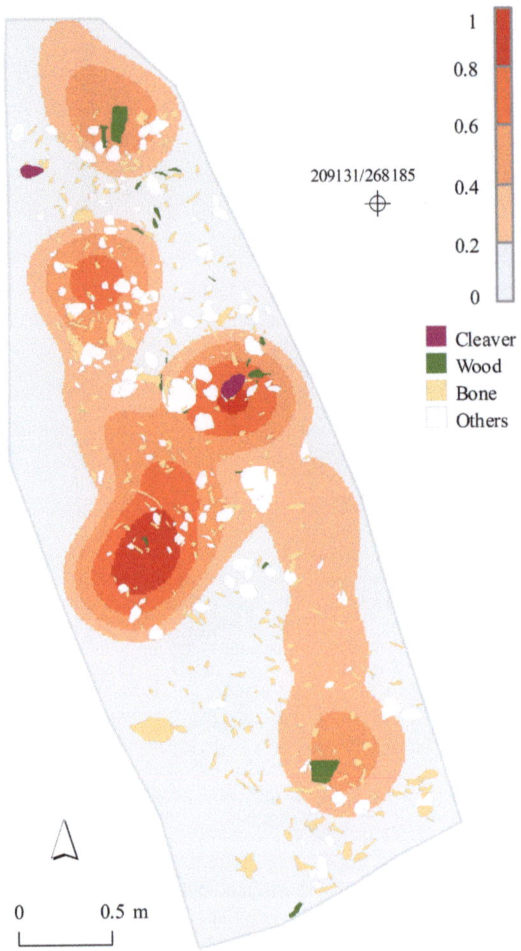

Fig. 4.3 Layer V-6. The field map superimposed on the kernel density map of burned flint microartifacts

(Rabinovich et al. 2008). The occurrence of all the major skeletal elements of *Dama* in this area, and the fact that they display evidence of all types of hominin-induced damage, suggested that "… the entire sequence of carcass processing took place in situ" (Rabinovich et al. 2008:143). This systematic butchering sequence was repeatedly practiced, suggesting that the hominins of GBY "… possessed detailed knowledge of their prey's anatomy and exploited game in an efficient, systematic manner" (Rabinovich et al. 2008:143). Interestingly, the similarities that emerged from the comparison between the *Dama* exploitation patterns at GBY with those of *Dama mesopotamica* from the Upper Paleolithic site of Hayonim Cave suggest that the expertise displayed by the GBY hominins in processing the *Dama* carcasses is comparable to that of the anatomically modern humans of Hayonim (Rabinovich et al. 2008).

The spatial distribution of the bone assemblage from Layer V-6 was examined to inquire into the possible role of fire in relation to animal-processing activities.

The large bone assemblage of Layer V-6 comprises 1,869 items (R. Rabinovich 2007, personal communication). Spatial data were available for 1,795 bones, which were spatially plotted and examined. The spatial distribution of the bone assemblage was examined according to several parameters (e.g., species definition, cut marks, breakage, and gnawing marks). No clustering was observed for any of the examined parameters, suggesting that the bones are randomly distributed throughout the excavated surface. In addition, the area surrounding the two concentrations of burned flint microartifacts (i.e., an oval area of 1.55 m^2) was examined to ascertain the frequencies of various faunal elements in that area. As with the general distribution of the bone assemblage, this area does not exhibit significant patterns of difference for the characteristics of the bones (Table 4.4).

Table 4.4 Layer V-6: characteristics of the general bone assemblage and of the bones in proximity to the concentrations of burned flint microartifacts (C)

	Total assemblage		In proximity to C	
	N = 1,795		N = 475	
	N	%	N	%
Cut marks	73	4.06	13	2.73
Gnawing marks	58	3.23	16	3.36
Elephant	20	1.11	7	1.47
Dama species	232	12.92	55	11.57

Of the large number of bones of Layer V-6, 332 are represented in the field map. The field map clearly illustrates the dense appearance of the faunal assemblage of Layer V-6. In addition, it seems that larger bones occur in the southern part of the excavated surface (Fig. 4.3).

In sum, the relatively small assemblage of burned flint microartifacts (N = 83) of Layer V-6 displays two adjacent high-density concentrations in the central part of the excavated surface. This area is spatially associated with one of the two concentrations of basalt microartifacts, suggesting that knapping of basalt was partly associated with the presumed location of the hearths. In addition, the analysis of the basalt component suggested that the concentrations of basalt microartifacts are spatially associated with basalt macroartifacts, particularly cleavers. This is markedly different from the distribution of flint microartifacts and macroartifacts, which are scattered throughout the exposed surface.

The faunal assemblage, which attests to the carcass processing of several species (particularly *Dama*) as well as marrow consumption, seems to be randomly distributed throughout the exposed surface. No clustering of specific species or of particular features (e.g., human- or animal-induced damage) was observed.

Although the faunal assemblage of Layer V-6 is suggestive of in situ processing of *Dama* (Rabinovich et al. 2008), the *Dama* bones, like the general assemblage, seem to be randomly distributed. Various parameters (discussed above) suggest that Layer V-6 was rapidly covered and sealed. Yet, unlike the lithic assemblage, which preserves its original spatial patterns, the bone assemblage seems to have been spatially reorganized.

Two agents can be considered as potential disturbers of the spatial configuration of the bones of Layer V-6: carnivore and human involvement. Considering the small percentage of carnivore gnawing marks on the bones of Layer V-6, it seems more likely that hominin activity is the agent responsible for the spatial rearrangement of the bones. The field map suggested that larger faunal elements are more common in the southern part of the excavated surface (Fig. 4.3), although these observations are not sufficient evidence for the existence of a dumping or tossing area of bones here.

4.3.2 The Elephant-Butchering Site of Layer II-6 L-1

The archaeological assemblage of Layer II-6 L-1 yielded a variety of lithic, faunal, and botanical remains. The latter comprise seeds, fruits, bark, wooden logs, and wood fragments. Some 83% of the wood segments of this layer (142 out of 171) were botanically identified and include ash (*Fraxinus syriaca*), olive (*Olea europaea*), Atlantic terebinth (*Pistacia atlantica*), oak (*Quercus* sp.), and willow (*Salix* sp.) (Goren-Inbar et al. 2002b: Table 13). The lithic assemblage incorporates giant basalt cores (Madsen and Goren-Inbar 2004) and their end-products, namely basalt handaxes and cleavers (Goren-Inbar and Saragusti 1996), as well as a variety of cores and tools of various raw materials (i.e., basalt, flint, and limestone).

The unique discovery of this layer is the skull of a straight-tusked elephant (*Palaeoloxodon antiquus*), which was found closely associated with stone and wood items (Fig. 4.4). The specimen was found in a good state of preservation and comprises most of the facial part of an adult female elephant cranium (Goren-Inbar et al. 1994).

Interestingly, the entire palatal and basicranial region of the skull had been removed, and the remaining surface is totally crushed. The skull exhibits a damaged area, ca. 30 cm long by 6 cm across, just below the nasal opening; the bone surface in this area has been removed by percussion to a depth of 3 cm. Thousands of tiny bone fragments were found embedded in the sediment adhering to and in direct proximity to the skull. About half of these fragments have the thin-walled appearance that is typical of the inner lining of the brain case and sinuses of an elephant skull, and very likely belong to it (Goren-Inbar et al. 1994). In addition, small fragments of elephant molar and a fragment of an elephant tusk were found in close proximity to the skull. Other, larger skull and tusk fragments were found in the area surrounding the skull.

The skull was recovered with the base facing up, laid on a large basalt core and a 1.5 m long log of oak (the hardest wood in the assemblage). Some seven basalt bifaces were found in the area surrounding the skull. The spatial configuration of the battered elephant skull, the wooden log, the basalt core and the basalt tools does not point to a uniform orientation. Furthermore, the good preservation state of the cranial surface,

Fig. 4.4 Layer II-6 L-1. (**a**) The exposed living floor during field work and (**b**) the elephant skull after retrieval from the field

the lake-side sedimentary context, and the co-occurrence of large items and small fragments of bone and stones suggest that fluvial transport is an unlikely explanation for the presence of the elephant skull (Goren-Inbar et al. 1994).

The findings of Layer II-6 L-1 are interpreted as the remnants of an in situ activity that involved the processing of the elephant skull. The battering damage on the skull and the numerous occipital and alveolar fragments (found in their "correct" anatomical positions relative to the cranium) suggest that breakage occurred while the skull was still exposed on the surface and that the skull was treated to extract the highly nutritious brain (Goren-Inbar et al. 1994). This procedure may have involved the use of the large basalt core, deliberately positioned under the skull in order to prevent it from sinking into the mud, as well as the long log, which may have been used as a lever to maneuver the skull and turn its basal side up (Goren-Inbar et al. 1994). The involvement of hominins in processing the skull is clear, considering the battering damage and the position of the skull and skull fragments, although direct evidence that the hominins hunted the elephant is more difficult to identify. However, the various lines of evidence discussed above are suggestive that "... the observed arrangement of finds actually reflects the terminal, dismemberment stage of the hunt ... the killing stage had been planned in advance to take place at the location most convenient for the hunters. As it seems against the nature of the wounded animal to have struggled purposefully into the midst of a hominid occupation zone ... towards the end of the hunt it might have been driven there by the Acheulian hunters" (Goren-Inbar et al. 1994:109).

In the framework of this study, we examine the role of fire within this particular activity. As previously discussed (Section 3.8), analyses of the spatial distribution of the burned flint microartifacts (N = 754) concluded that two clusters of burned material are recorded in Layer II-6 L-1 and that these clusters are remnants of anthropogenic-controlled fire in the form of open hearths. Examination of the spatial association of these hearths with other components of the occupation horizon illustrates that the elephant skull is located between the two hearths (Fig. 4.5), suggesting that the use of fire was involved in the processing of the elephant. In addition, several observations can be drawn from the spatial configuration of the finds from Layer II-6 L-1:

- The exposed archaeological horizon illustrates the distribution of various categories of the archaeological assemblage of Layer II-6 L-1 (Fig. 4.5). Despite the overall density of finds on the living floor, the vicinity of the phantom hearths is not as dense as the surrounding area, particularly the northern hearth, where a clear void is observed.

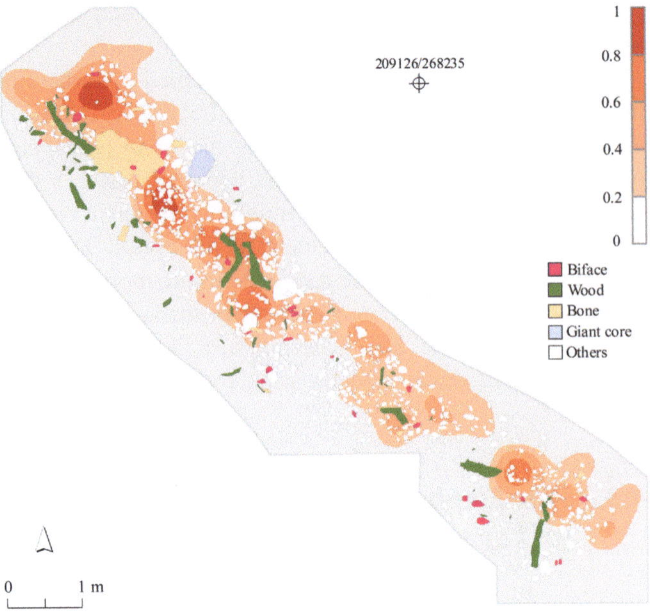

Fig. 4.5 Layer II-6 L-1. The field map superimposed on the kernel density map of burned flint microartifacts

- The distribution of wood pieces throughout the archaeological horizon (Fig. 4.5) illustrates the large amount of preserved *unburned* wood pieces. Layer II-6 L-1 yielded an extensive assemblage of wood segments (N = 171), none of which was identified as burned (Goren-Inbar et al. 2002b). The presence of the unburned wood segments reinforces the assumption that the burned clusters represent remnants of human-controlled hearths.
- Bifacial tools appear to be scattered throughout the excavated surface. With the exception of five items (three of which are associated with the elephant skull), most bifaces fall outside the area of the phantom hearths (Fig. 4.5).

The distribution patterns of the archaeological material of Layer II-6 L-1 suggest a close proximity between the hearths and the battered elephant skull. This spatial association may imply that the processing of the skull, and perhaps the consumption of the animal's brain, involved the use of fire.

4.3.3 The Basalt Biface Workshop Site of Layer II-6 L-4

The excavation of Layer II-6 L-4 exposed an extremely dense concentration of bifacial tools, primarily basalt handaxes and cleavers (Fig. 4.6). These were found, together with other stone artifacts of various raw materials (basalt, flint, and limestone; see Table 3.13; Section 3.11), a relatively small assemblage of bones (N = 111; R. Rabinovich 2007, personal

Fig. 4.6 Layer II-6 L-4. View of the northern part of the excavated living floor during field work

communication), and organic material such as wood, bark, fruits, and seeds (Goren-Inbar and Saragusti 1996). Wood fragments from this layer (N = 46, of which 41 are taxonomically identified) include oak (*Quercus* sp.), pistachio (*Pistacia vera*), and ash (*Fraxinus syriaca*) (Goren-Inbar et al. 2002b:23, Table 4).

The bifacial tools are all patinated and exhibit variable degrees of weathering (i.e., from slight abrasion to complete exfoliation). Infrared spectrometry analysis conducted on the most degraded basalt artifacts suggested that, since their chemical composition is entirely clayey, these basalts apparently underwent in situ chemical weathering (Goren-Inbar and Saragusti 1996).

Detailed technological analysis suggested that most bifaces were made on large, wide basalt flakes of various techniques (e.g., Kombewa and Levallois), so that minimal retouch was required in order to modify the sharp active edges of the tools. The flakes on which the basalt handaxes and cleavers were modified were the subject of extensive knapping experiments (Madsen and Goren-Inbar 2004), which underlined the role of selection and transportation of good-quality basalt by the GBY hominins. These experiments suggested that, in view of the various size categories of the artifact classes and the under-representation of certain artifacts in the archaeological assemblages, it is likely that the flaking of heavy (i.e., giant) basalt cores commenced at the outcrop and that suitable large flakes were sorted and selected as blanks for future preparation of handaxes and cleavers (Madsen and Goren-Inbar 2004).

The detailed technological analyses of the bifacial component of Layer II-6 L-4 (Goren-Inbar and Saragusti 1996) further suggested that "... handaxes and cleavers are morphometrically very similar to each other, regardless of the specific technique by which they were modified" (Goren-Inbar and Saragusti 1996:24). Morphometric analyses, which applied a geometric model to the handaxe assemblage of Layer II-6 L-4, demonstrated that the artifacts' outlines display uniformity in "geometric shape" (Brande and Saragusti 1996). The fact that the bifacial component of Layer II-6 L-4 displays a high level of standardization and homogeneity in the size and morphology of the end-products is an indication of knapping expertise and may even be suggestive of an expert (i.e., individual) knapper.

In our attempt to examine the role of fire in various activities, the biface assemblage (as illustrated in the field map) has been superposed on the density map of the burned flint microartifacts (N = 1,406), where three concentrations of burned flint microartifacts are recorded (Fig. 3.40). The dense appearance of the bifacial tools is clearly illustrated in Fig. 4.7. Despite the general density, only a few bifaces occur within the high-density kernels of the burned flint microartifacts (Fig. 4.7). Apart from this, the biface assemblage does not exhibit particular patterns of distribution (e.g., clustering) in relation to the concentration of burned flint microartifacts and seems to be spread throughout the excavated surface.

The lithic assemblage of Layer II-6 L-4 does not exhibit characteristics of a regular workshop site. The abundance of basalt bifaces is not matched by the number of basalt waste products that is likely to be found in association with such a large amount of end-products (Goren-Inbar and Sharon 2006). Indeed, experimental studies have suggested that the flaking of heavy (i.e., giant) basalt cores commenced at the outcrop and that suitable large flakes were sorted and selected as blanks for future preparation of handaxes and cleavers such as those of Layer II-6 L-4 (Madsen and Goren-Inbar 2004). The archaeological and experimental data suggest that the bifaces of Layer II-6 L-4 may represent evidence of a stocking site to which bifacial tools were transported as finished products.

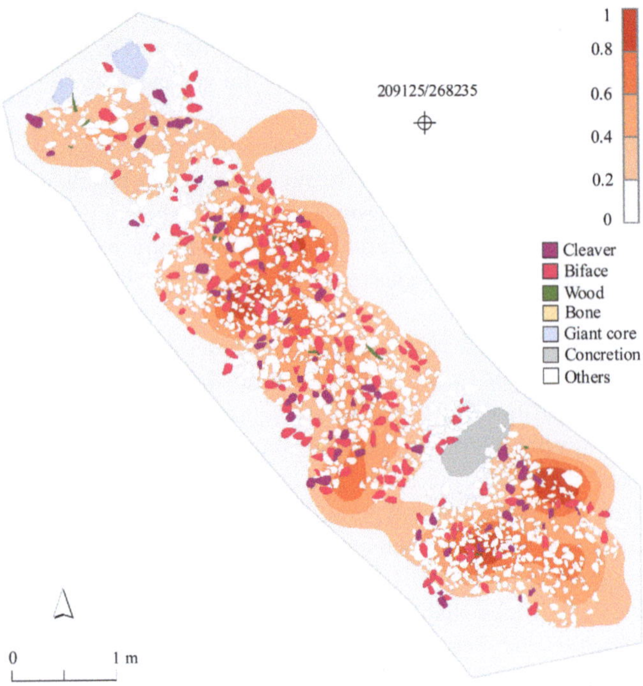

Fig. 4.7 Layer II-6 L-4. The field map superimposed on the kernel density map of burned flint microartifacts

4.3.4 Humans and Fire at GBY: Summary

A detailed presentation of the spatial configuration of each of the archaeological horizons of GBY is beyond the scope of this study. However, in order to illustrate the variety of activities associated with fire, lithic and faunal assemblages from several archaeological layers have been spatially analyzed.

The opening paragraph of Section 4.3 raised a question with regard to the role of fire: is the use of fire associated with particular tasks or is it an integral part of the cultural repertoire, regardless of the type of activity carried out in each of the occupational levels? The three archaeological layers presented above demonstrate that fire occurs in association with a variety of activities, and is certainly not limited to a particular task.

In Layer V-6, the concentrations of flint and basalt microartifacts do not coincide and exhibit different distribution patterns, suggesting that the knapping of flint was carried out throughout the occupational surface while the knapping of basalt was confined to the northern part of the exposed surface. One of the two high-density concentrations of basalt microartifacts is in close vicinity to a high-density concentration of burned flint microartifacts. This may suggest that the knapping of basalt was carried out close to the presumed location of the hearth.

While in Layer V-6 the knapping of basalt can be spatially associated with the presumed hearth location, in Layer II-6 L-4 no association was found between the high-density concentrations of burned flint microartifacts and the numerous basalt handaxes and cleavers. The particularly dense appearance of these tools may represent stocking of basalt bifaces, which is not spatially associated with the use of fire.

In Layer V-6 the rich and dense faunal assemblage did not exhibit any spatial association with the concentrations of burned flint microartifacts. Thus the presumed hearth locations are not related to a particular species or to a particular feature of carcass processing (e.g., cut marks, percussion marks). Such an association, however, is illustrated at the elephant butchering site of Layer II-6 L-1. There, the battered elephant skull, which exhibits various damage features suggesting that the skull was processed for the purpose of extracting the highly nutritious brain, was found between the two concentrations of burned flint microartifacts. The position of the skull in that area suggests that the processing of the skull, as well as the consumption of the elephant's brain, may have involved the use of fire.

In a previous study of pitted anvils from GBY it has been demonstrated that the co-occurrence of these pitted stones and of edible nuts remains is suggestive of the processing of nuts by the Acheulian hominins at the site (Goren-Inbar et al. 2002a). Pitted anvils, which occur throughout the stratigraphic sequence at GBY, are cores, blocks, slabs, and flakes made on basalt and limestone that exhibit pits of various quantities, sizes, and depths (Goren-Inbar et al. 2002a). In the framework of the current study, spatial analyses suggested that in several archaeological layers these pitted stones occur in distinct clusters.

Interestingly, in Layer II-6 L-2 seven of the eight pitted stones recorded in the layer are found within and in close vicinity to the concentration of burned flint microartifacts (Fig. 4.8). The field map of this layer illustrates that the location of the phantom hearth is also the area in which unburned wood segments and the bulk of the bifacial tools are recorded (Fig. 4.8). Another interesting phenomenon illustrated in Layer II-6 L-2 is the fact that the largest basalt item exposed on the living floor is recorded in close vicinity to the phantom hearth (Fig. 4.8). The compilation of these data suggests that in Layer II-6 L-2 the hearth served as a focal point of activities. The clear spatial association between the hearth and the pitted stones is suggestive that the processing of nuts may have involved the use of fire.

Similar patterns emerge from the distribution of archaeological finds from Layer II-6 L-6, whereby the largest basalt items are recorded in close vicinity to the location of the phantom hearth (Fig. 4.9). The distribution patterns of pitted stones in Layer II-6 L-6 (N = 23) exhibit three clusters; one of these includes eight items and is recorded within and in the immediate vicinity of the phantom hearth (Fig. 4.9). Furthermore, bifacial tools, particularly handaxes, are not recorded in the area of the phantom hearth but clearly are

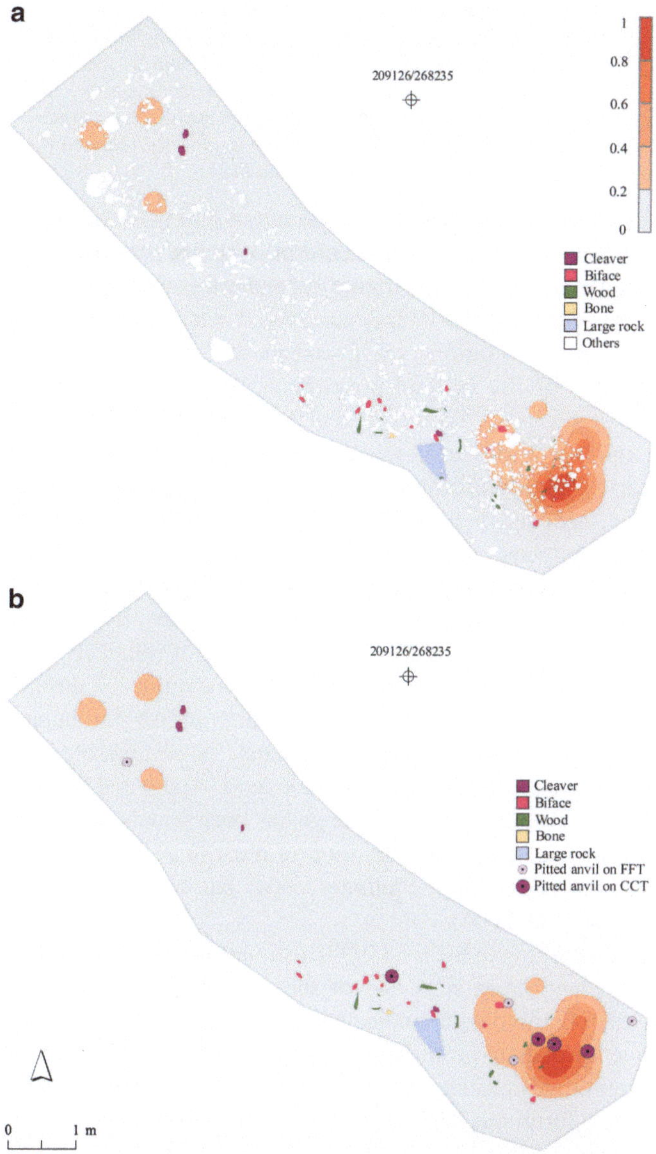

Fig. 4.8 Layer II-6 L-2. (**a**) The complete field map superimposed on the kernel density map of burned flint microartifacts and (**b**) selected items of the field map (largest basalt slab, bones, wood, handaxes, and cleavers) and the pitted anvils superimposed on the kernel density map of burned flint microartifacts

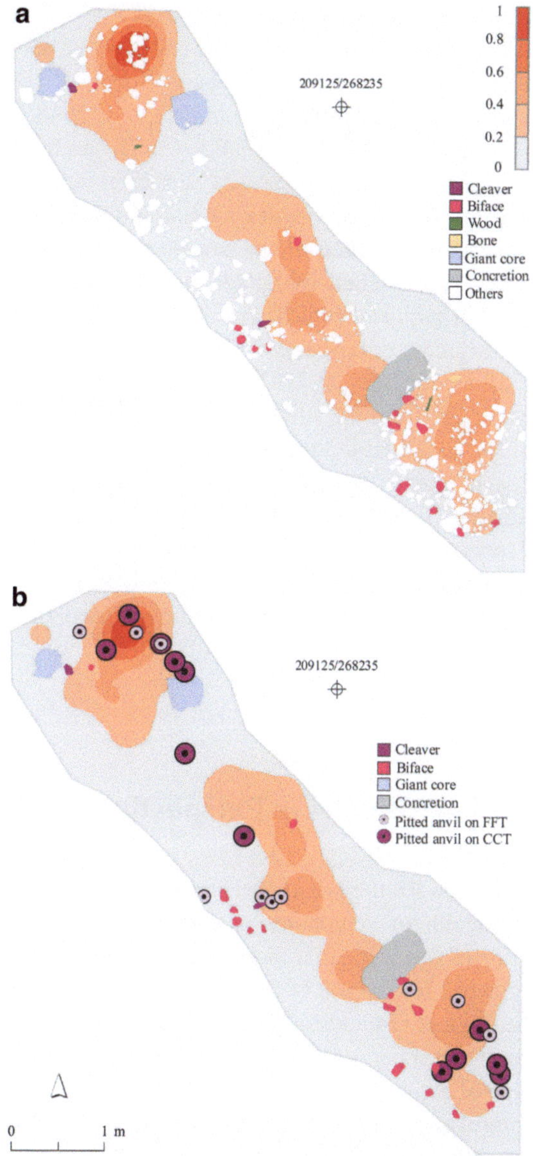

Fig. 4.9 Layer II-6 L-6. (**a**) The complete field map superimposed on the kernel density map of burned flint microartifacts and (**b**) selected items of the field map (largest basalt cores and bifacial tools) and the pitted anvils superimposed on the kernel density map of burned flint microartifacts

spatially associated with the other two concentrations of pitted stones (Fig. 4.9).

In sum, the spatial analyses of different components of the archaeological record at GBY have uncovered complex and variable patterns of distribution. These illustrate that the archaeological record has preserved indisputable evidence for spatial patterning of a variety of activities (e.g., stone knapping, carcass processing, nut cracking). The extremely complex and diversified issue of spatial patterning of activities is beyond the scope of this study. However, in the framework of this study the archaeological evidence has clearly demonstrated that the use of fire at GBY was an integral part of the Acheulian cultural repertoire and that fire was used by the hominins throughout the entire occupational sequence. In several cases we have been able to establish a spatial association between the hearth location and particular activities (e.g., knapping of basalt in Layer V-6, processing of the elephant skull in Layer II-6 L-1, processing of nuts in Layers II-6 L-2 and II-6 L-6).

In other cases, however, we have encountered patterns of distribution that are not spatially associated with fire (e.g., stocking of basalt bifaces in Layer II-6 L-4, butchering of *Dama* carcasses in Layer V-6), or patterns that seem to be spatially associated with the hearth location (e.g., the occurrence of large basalt items in proximity to the hearths in Layers II-6 L-2 and Layer II-6 L-6). The archaeological record at GBY clearly preserves further information that may shed light on the complex spatial patterning of activities carried out in each of the archaeological horizons at the site. A comprehensive understanding of this complex system will require detailed spatial analyses of the various archaeological components of each of the hominin occupation episodes at GBY. The study of fire-damaged microartifacts from the site has established evidence for the use of fire and enabled the reconstruction of the locations of ancient hearths throughout the occupational sequence. In addition, it has been suggested that the use of fire at GBY should be considered a major component in any study of spatial patterning, as it most likely served as a focal point for a variety of activities carried out by the Acheulian hominins.

4.4 Paleolithic Fire in Perspective

The evidence for the use of fire presented in this study adds to the wealth of accumulating data on the activities and behavior of the Acheulian hominins of GBY. The fact that the evidence for the use of fire is continuous throughout the archaeological sequence at GBY implies that the use of fire was an integral part of their cultural repertoire, regardless of the type of activity.

Control over fire provided the hominins of GBY with a remarkably powerful tool. In the framework of this study we have presented evidence that spatially associates the use of fire with a variety of activities (e.g., stone knapping, animal processing, and nut cracking; see Section 4.3). Moreover, it is likely not only that fire was incorporated into these everyday activities of the GBY hominins, but also that it had far-reaching implications for their lives that cannot be detected in the archaeological residues.

The advantages conferred on humans by the control and use of fire are diverse and include warmth and light, protection from predators, the exploitation of a new range of foods, the emergence of new technologies that require fire use, and a variety of behavioral and social benefits. The following discussion attempts to explore various aspects of the human mastery of the use of fire. Though it is clear that we cannot yet directly ascribe the full scope of these advantages to the hominins of GBY, we will attempt to shed light on the options that fire may have provided for these ancient hominins.

> And then the sun taught them to cook food and soften it by the heat of flame, since they saw many things among the fields grow mellow, vanquished by the lashing of his rays and by the heat. And day by day those who excelled in understanding and were strong in mind showed them more and more how to change their former life and livelihood for new habits and for fire. (Lucretius, *On the Nature of Things* Book V: 1102–1105)

These archaic thoughts on the nature of things encompass the realization that human control over fire was a cultural point of no return. Furthermore, it emphasizes the role of fire in food preparation, which is considered a major evolutionary incentive. Evolutionary adjustments that may be related to the use of fire in food preparation are often attributed to *Homo erectus* and contemporaneous hominin taxa. *H. erectus* exhibits various anatomical traits that distinguish it from earlier hominins, including an overall increase in body size, reduced body size dimorphism, smaller molars and jaws, reduced prognathism, and increased brain size. Aiello and Wheeler (1995) have established a hypothesis for this evolutionary shift. Their "expensive-tissue hypothesis" suggests that the metabolic requirements of relatively large brains are offset by a corresponding reduction of the gut. Gut size is highly correlated with diet and in particular a small gut (characteristic of humans) is compatible only with high-quality, easily digestible food (Aiello and Wheeler 1995). This evolutionary change has been widely associated with an increased consumption of meat (e.g., Milton 1999; Kaplan et al. 2000; Plummer 2004; but see discussion in Ungar et al. 2006).

According to Milton (1999), estimations of early hominin body and brain size, and of the anatomy and kinetics of extant hominoid gut, indicate that the most expedient dietary avenue open to proto-humans was increasing reliance on the intentional consumption of animal matter on a routine rather than a fortuitous basis. This dietary shift, from lower-quality plant food to higher-quality meat, became possible through technological innovations that enabled nonsomatic digestion (i.e., food preparation) prior to food consumption. In humans, nonsomatic digestion frequently occurs before a food item is ever brought into contact with the mouth and gastrointestinal tract, a behavior that could ultimately have affected human gut proportions (Milton 1999).

It has also been suggested (Wrangham et al. 1999) that certain anatomical traits of *H. erectus* can be linked with a dietary change that involved such nonsomatic digestion, i.e., cooking. Cooking increases energy intake substantially more than substituting meat for plant material (Wrangham et al. 1999). The fact that the consumption of cooked meat provides an energetic benefit over the consumption of raw meat has been recently supported by experimental studies (Boback et al. 2007); these experiments indicated that consumption of cooked meat decreases the cost of gastric digestion and reduces the investment of time and effort in chewing (Boback et al. 2007). Cooking improves the digestibility of food by eliminating physical barriers (e.g., thick skins and husks) and by changing the structure of molecules as in proteins and starches. In addition, the use of fire in food preparation by cooking, roasting, or smoking makes food more durable and is a means of detoxifying certain poisonous foodstuffs, thus substantially broadening the range of edible species (e.g., Stahl 1984; Wandsnider 1997). The ability to identify and avoid certain poisonous foods is physiologically expressed by a bitter taste. However, through the evolution of the human lineage some have lost this ability. Recently, the genes of bitter taste receptors of humans and other primates were cloned and their sequences analyzed; these studies have suggested that the loss of constraint occurred in two phases, an early one that affected the ancestry of both humans and chimpanzees and another one at approximately 0.75 Ma that affected the hominin lineage alone (Wang et al. 2004). Since cooking detoxifies poisonous food, by consuming cooked food hominins reduced their exposure to toxins and as a result reduced the selective pressure of human bitter taste genes (Wang et al. 2004; Meyerhof 2005). Interestingly, the estimate of the starting time (i.e., 0.75 Ma) of the complete functional relaxation in human bitter taste genes coincides with the age of the Acheulian site of GBY, where controlled use of fire is recorded.

As cooked foods (of both animal and plant origin) require less extensive digestion than raw foods, the adoption of cooking may have influenced the morphology of dentition and the intestine, reducing the size of teeth and gut (Aiello and Wheeler 1995). Wrangham et al. (1999) have extended this argument, emphasizing the role of fire, and particularly cooking, in the consumption of *plant* food by early hominins; thus it was not necessarily the consumption of meat or cooked meat that triggered the dietary and evolutionary shift, but rather the cooking of *plant* materials (Wrangham et al. 1999). It is noteworthy that a virtually intuitive preference for cooked plant foods has been recorded for a group of chimpanzees following a bush fire: "… I was amazed to see Tina sorting through the ashes at the base of an afzalia tree and picking up the charred-looking seeds. Once cooked, they crumbled easily between strong molars … the chimps loved them …" (Brewer 1978:232). The hypothesis that great apes perceive properties of cooked food and prefer them to those of raw food has recently been supported by an experiment showing that several populations of captive apes prefer their food, from tubers to meat, cooked rather than raw (Wobber et al. 2008). These experiments actually imply that the preference for cooked foods is a pre-adaptation, suggesting that when cooking was first adopted hominins would have readily preferred their food cooked (Wobber et al. 2008).

Since the cooking of plant foods increases energy availability, it may have also enabled an intensification of the high-risk activity of hunting (Wrangham et al. 1999). Hunting is yet another example of the benefits that fire may have brought to ancient hominins; particularly, the use of fire in hunting may have further expanded the dietary change provided by fire. The exclusively human control over fire has affected the relationship between human beings and the world they live in, including their relationship with other animals, as clearly illustrated by the course of human supremacy over other animals and the differentiation in power and conduct between hominins and other large mammals (Goudsblom 1992). This human supremacy is particularly expressed by hunting. However, apart from the presence and character of fossil faunal evidence, archaeological evidence for hunting is limited, and recognizable indications of the use of fire in hunting are even scarcer. Nevertheless, a variety of archaeological, and particularly ethnographic, data suggest that humans have used fire in order to chase and hunt game (e.g., Day 1953; Grayson 2001; Rolland 2004). Furthermore, the burning of particular habitats enables humans to drive animals and thus enhance their hunting ability. Recurrent burning allows humans to control these habitats by improving the nutritional quality of the forage supplies available for the animals thus ensuring availability of the hunt in these habitats (e.g., Sauer 1956, 1947; Lewis 1972; Mellars 1976; Naveh 1990; Marlowe 2005; Fairbairn et al. 2006).

The use of fire eventually led humans to experiment with this powerful tool and to give rise to a variety of technological innovations. Fire-using technologies, or pyrotechnologies, occupy a major place in the attempts to model and alter natural resources and materials: "… we cannot help marveling that there is almost nothing that is not brought to a finished state by means of fire. Fire takes this or that sand, and melts it, according to the locality, into glass, silver, cinnabar, lead of one kind or another, pigments or drugs. It is fire that smelts ore into copper, fire that produces iron and also tempers it, fire that purifies gold… Fire is a vast, unruly element, and one which causes us to doubt whether it is more a destructive or a creative force" (Pliny, *Natural History* Book XXXVI:200–201).

The foundations of such pyrotechnologies are rooted in our prehistoric past and can be traced in the use of fire for quarrying and mining (e.g., Forbes 1963; Shepherd 1994; Voytek 1997; Ambret 2002) or for splitting boulders/nodules of raw material (see references in Rolland 2004: Table 1), methods that may have been used as early as the Lower Paleolithic. Oakley (1956), for example, described a Late Acheulian sandstone factory site in South Africa where large and thermally altered flakes were found. Oakley further reported that "… Acheulian man obtained the flakes by the method of "fire-setting", that is to say burning fires around the blocks of sandstone and then dashing water on the heated rock to cause exfoliation" (Oakley 1956:45).

In addition to the procurement of raw materials through quarrying or mining, fire is known to have contributed to the process of manufacturing and modifying stone tools, either through the controlled heating of flint in order to improve its knapping quality (e.g., Crabtree and Butler 1964; Inizan et al. 1975; Price et al. 1982; Purdy 1982) or through the hafting of stone tools with adhesives that require the use of fire in their preparation (e.g., Inizan 1976; Tankersley 1994; Boëda et al. 1996). The innovative art of using fire in the preparation of adhesives enabled ancient tool-makers to replace broken or damaged tools by melting the adhesive, retooling the tool, and drying the adhesive again. Recent experimental studies of adhesives have suggested that the use of fire in the initial stages of adhesive manufacture is essential, as it takes off excess moisture and makes the adhesive more elastic (Wadley 2005).

Another fire-using technology that is embedded in our prehistoric past is the use of fire to harden wood. Although ethnographic evidence (e.g., Marlowe 2005) demonstrates that hunter-gatherers use fire to harden digging sticks, archaeological evidence of this innovation is scarce, since organic remains are rarely preserved through time. Nonetheless, wooden spears that may have been hardened by fire are reported from several prehistoric sites, such as Clacton-on-Sea (Howell 1966; White 2000), *Kalambo Falls* (Clark 2001), Lehringen (Movius 1950; Oakley 1956; Thieme and Viel 1985), *Schöningen* (Thieme 1997), and *Torralba* (Howell 1966). Interestingly, at Lehringen the fire-hardened spear was found thrust between the ribs of a straight-tusked elephant (Thieme and Viel 1985). Such findings suggest that fire was utilized to harden wood and that wooden spears were used for hunting during early stages of the Middle Pleistocene.

This short review of fire-using technologies demonstrates that obtaining control over fire enabled ancient hominins to initiate the continuous process of modifying and transforming the resources provided by nature and that "… both experimental curiosity and the spirit of invention were as animated then as they are today" (Rudgley 1999:142).

While discussing fire-using technologies, we should bear in mind the fundamental qualities of fire that enabled its manipulation in a wealth of activities such as cooking, quarrying, hafting, or hunting. Fire is first and foremost a source of heat and light, and as such provided humans with warmth and comfort that modified and improved their living conditions. Fire enabled humans to inhabit new niches (e.g., caves and rock shelters), as it provided the necessary illumination as well as protection against carnivores and pests (e.g., insects). Furthermore, fire, and the warmth and light it offered, expanded the range of activities, enabling these to take place after sundown: "… clearly, the fire furnished a setting around which a group could gather and engage in common activities well into the night … needless to say, the power exercised through fire was essentially social – that is, it could be sustained only by a group … It was simply impossible to keep a fire burning for long without at least some social cooperation and division of labour in order to guard it and fuel it" (Goudsblom 1992:40).

The image of a group sitting around a smoldering fire depicts the significance granted to fire as a socially bonding resource. The fact that humans carried out a variety of hearth-related communal activities may eventually have led to the development of complex societies: "… learning to control fire was, and is, a form of civilization. Because humans have tamed fire and incorporated it into their own societies, these societies have become more complex and they themselves have become more civilized" (Goudsblom 1992:3). Goudsblom uses "civilization" in a sense that is not exclusive to modern societies, but rather describes a dynamic process of learning through which humans were able to change their habits and culture. Similarly, Perlès (1977) views the use of fire as a "mental progress" rather than a technical progress that required new forms of social organization: "… Il nous semble de toute façon que, du point de vue de l'évolution humaine, le changement le plus important a été le passage de la non-utilisation du feu à l'utilisation du feu … en effet, la découverte de l'utilisation du feu suppose un progrès psychique et non technique …" (Perlès 1977:30).

Fire must have played a significant social role for ancient humans and influenced social organization and social structure. This may be related to the fact that, along with the advantages that fire brought to humans, it also made them dependent on its benefits as they became more and more vulnerable to its loss. This must have meant that a constant supply of fuel was regularly collected and that the knowledge of fire-making or fire-keeping was protected. Such imperatives likely required the cooperation of a group of humans, who strengthened their group ties and group reliance through their shared interest in fire-keeping and fire-using. Ethnographic evidence (e.g., Frazer 1930; Stewart 1956; Blainey 1975) suggests that the early stages of human use of fire may have involved the social role of "fire-keeper": "The capture of fire brought with it the duty of keeping it 'alive', as we still say. Ages may have elapsed before the art of making fire was discovered, and then the making of new fire was not easy. Continuous possession of fire is a theme of ancient religions as sacred or eternal fire, in the care of female attendants …" (Sauer 1961a:261). As humans gained the knowledge of fire-making, they may have required a different social role – the "fire-maker". As stated by Sauer (1961a:261), the ethnographic record illustrates that these social roles are often gender-associated (e.g., Frazer 1930; Raum 1973).

The social and behavioral adaptations that emerged following the controlled use of fire are closely related to the use of fire in food preparation, which occupied the opening part of this discussion. It is a consensual view that the adoption of new dietary avenues, and in particular the use of fire in food preparation, has directly influenced social bonds: "Like the control of fire, cooking is an element of culture.

It has to be learned, and this learning is done in groups. It demands a certain amount of division of labour and mutual cooperation, and also, at the individual level, attention and patience … and already at a much earlier stage, the social coordination and the individual discipline acquired as a result of cooking might have useful spin-off effects in other activities as well" (Goudsblom 1992:35). Similarly, Perlès has noted that: "… Sans aller aussi loin dans un domaine où tout est conjecture, tout porte à croire effectivement que la cuisson de la nourriture a pu avoir une influence sur le mode de pensée de l'homme. Mais il est un domaine où nous pouvons être certains que l'introduction de la cuisson des aliments a eu des conséquences réelles: l'organisation du cycle des activités quotidiennes. En effet, la cuisson entraîne nécessairement une importante dépense de temps: temps d'approvisionnement en combustible – temps de préparation du foyer – temps de préparation des aliments – temps de cuisson proprement dite. L'organisation des activités du groupe est donc modifiée. De plus, il est probable qu'à ce niveau apparaît une spécialisation des members du groupe, certains étant plus particulièrement préposés à ces tâches" (Perlès 1977:101).

By incorporating fire into their dietary habits, humans clearly extended their diet with new substances that in their raw state would have been difficult to consume or even toxic and harmful; furthermore, the cooking of food led human groups to develop particular eating habits, by which they could distinguish themselves from other animals and also from each other. Moreover, the use of fire in food preparation involved communal behavior and occupied, as it still does, a significant role in human activities and gatherings. In the view of Wrangham et al. "… cooking of both animal and plant foods likely involved central-place foraging with delayed consumption of food, which brought otherwise dispersed plant foods into a category previously proposed for hunted animal products: packages amenable to sharing …" (Wrangham et al. 1999:568).

This view brings us to the concluding issue of this discussion: the role of fire in the site function and mobility strategies of early hominins. The Plio-Pleistocene deposits of the FLK Zinj floor at Olduvai Gorge (Leakey 1971) generated a new interpretation, as living floors or home bases, of the observed co-occurrence of animal bones of various species and sizes with dense scatters of stone tools. The term "home base" was developed soon after by Glynn Isaac into the "food-sharing hypothesis" or "central-place foraging hypothesis" (e.g., Isaac 1978, 1983) and revolutionized our interpretation of ancient hominins' behavior, initiating an ongoing debate (e.g., Binford 1984; Potts 1984, 1994; Bunn and Kroll 1986; Sept 1992; Rose and Marshall 1996). The debate derives from the fact that the "home base hypothesis" attributed various adaptive behaviors to the ancient hominins. The use of a base camp, or a central place to which a variety of resources are brought on a daily basis, is also associated with delayed consumption of the food and with sexual division of labor in which males hunt or scavenge and females gather plant foods, with all resources being transported to the base camp where food sharing takes place.

Attributing to early hominins the ability to use and control fire is another factor that corresponds with the concept of the home base and its related adaptations. Fire is often associated with a perception of home (e.g., Galanidou 1997; Binford 1998; Rolland 2004, 2000) and it was probably Sauer (1947) who was the first to link home and fire. In his reflections on the culture of *H. erectus*, Sauer noted: "The presence of hearths shows that these folk lived with some degree of permanence at particular places. The number of hearths suggests that they may have lived in family groups, each accustomed to collect at its own place (hearths = home). There is nothing in this picture to indicate aimless wanderers, the mythical man pack or horde, drifting freely" (Sauer 1947: 158–159). Half a century later, these notions have been revived in the context of home bases, suggesting that fire occupied a significant role in the emergence of these residential localities. According to Rolland, a regular use of fire appeared as a discrete or "punctuated" event around 0.4 Ma and contributed to "… a major organizational shift in ancient hominid settlement and land use systems, expressed by 'home bases' " (Rolland 2004:270).

Indeed, the manipulation of fire likely involved changes in human mobility patterns and enabled hominins to occupy a particular location for prolonged periods of time. There, a variety of behavioral and social adaptations would have evolved, taking part in the process of human "civilization". However, this study has presented evidence that this step in our evolution occurred during a much earlier stage, within the Acheulian cultural technocomplex. The evidence for the use of fire and its associated activities through the sequence of ca. 0.1 Ma at GBY joins a variety of indications of hominin activities at the site. These activities encompassed diverse tasks (e.g., raw material procurement, stone knapping, animal butchering, processing of plant foods) that likely required social organization and the cooperation of several group members in the framework of a "home base", where the various lithic, plant, and animal resources were collected, communally processed, and used in the light of their domestic fire.

The manipulation of fire has led to dramatic changes in human diet, technology, evolution, and behavior. Beyond the clear functional advantages it provides, the domestication of fire has separated us from the animal world. While other animals are capable of various degrees of communication and "language" and some are able to make and use tools made of wood or stone, the ability to make and control fire is exclusively human. Thus, this stage in our evolution should primarily be considered a cultural and behavioral revolution, as more than anything else, the control of fire has made us human.

4.5 Conclusions

Since the manipulation of fire provided major advantages for our ancestors, attempts to establish evidence for the early use of fire have occupied a variety of studies. Nevertheless, the question of *when* humans obtained and controlled fire has remained open. The presence and spatial clustering of burned flint items at GBY offer a unique opportunity to explore this important issue and provide substantial evidence for controlled use of fire as early as 0.79 Ma.

The central assumption of this study is based on a variety of ethnographic, archaeological, and ethnoarchaeological studies, which emphasize the fact that small burned items are likely to preserve their original spatial configuration and serve as spatial indicators for the location of ancient hearths (Section 2.2). Thus, we have suggested that the possible use of fire by the Acheulian hominins at GBY may be revealed through spatial analyses of the burned flint microartifacts, and that potential concentrations of these should be considered remnants of hearths.

The results of this study (Chapter 3) have demonstrated that burned flint microartifacts and macroartifacts occur throughout the occupational sequence of GBY, exhibiting varying patterns of distribution. Furthermore, the burned flint microartifacts are never found uniformly distributed throughout the excavated surface, so that dense concentrations are recorded in each of the analyzed archaeological layers.

This study has considered the possibility that the burning of flint items at GBY may be the outcome of natural wildfires. In order to explore this possibility, the different factors involved in the complex process of fire ignition, combustion, and behavior were discussed. That discussion (Section 4.1) further examined the required conditions in which a natural fire may have ignited and spread in the particular environmental setting of GBY. The combination of the ecological data, the various environmental characteristics of the GBY sedimentological setting, and the archaeological record permits the firm rejection of the possibility that recurrent natural fires took place in the Acheulian lakeshore occupations of GBY.

As the scenario of a natural fire is unlikely, we have concluded that the concentrations of burned flint microartifacts in the different occupational surfaces of GBY are all the remnants of anthropogenic hearths. Comparison between the distribution patterns of the burned and unburned flint microartifacts in each of the occupational levels has suggested that these exhibit different degrees of overlapping, ranging from complete superposition to distinct differentiation. This variability is interpreted as reflecting the complexity of spatial patterning of activities. More specifically, in some cases flint knapping was carried out exclusively in vicinity to a hearth or hearths, so that the distributions of the burned and unburned flint microartifacts coincide, while in other cases flint knapping was not confined to the hearth area, so that the burned and unburned flint microartifacts are distributed in distinctively different clusters.

The spatial analyses presented in this study have emphasized statistically significant concentrations with particularly high densities of burned flint microartifacts. However, burned flint microartifacts are also recorded outside these clusters and we are required to consider their possible origin. First, it is probable that a small fraction of the burned microartifacts shifted from their original location – whether purposefully (e.g., as a result of tossing and clearing) or taphonomically (e.g., as a result of trampling). Another possibility is that several hearths of different intensities were in use within a single occupation level; in such a case, the "background" burned material that occurs outside the high-density clusters may in fact represent the remnants of hearths of lower intensity, which will in turn exhibit lower densities. It is also possible that the "background" burned material represents the margins of hearths, most of whose complete form (i.e., a concentration of burned flint microartifacts) occurs outside the excavated area.

While these questions and possibilities are all of great importance, the primary objective of this study was to differentiate the agents that may have introduced the burned flint artifacts and microartifacts to the archaeological levels at GBY. Nevertheless, it seems that the patterns revealed by this study not only suggest that hominins were using fire, but also present challenges of a higher order within the issue of early fire use that are beyond the scope of the present study.

Our interpretation of the spatial patterning of the burned and unburned flint microartifacts is (1) that the concentrations of burned flint microartifacts represent "phantom hearths", i.e., remnants of hominin use of fire in the form of hearths, and (2) that complex spatial patterning of activities is illustrated by the distribution of the flint microartifacts. These two conclusions are further supported by examination of the spatial association between the phantom hearths and other components of the archaeological assemblages. This examination has demonstrated that fire occurs in association with a variety of activities, including stone knapping, carcass processing, and nut cracking. These preliminary spatial analyses have uncovered diverse patterns (Section 4.3) and show that the archaeological record at GBY has preserved indisputable evidence for spatial patterning of activities. The fact that a variety of residues associated with different activities are distributed non-randomly provides unique evidence that the behavior of the Acheulian hominins of GBY was likewise not random across space. The spatial patterns illustrated in this study provide a significant contribution to the issue of the antiquity of spatial differentiation of activities, suggesting that such behavior existed at GBY as early as 0.79 Ma.

In addition, in outlining the research objectives of this study, we have addressed the issue of the magnitude of the use of fire throughout the various occupation episodes recorded at the site. More specifically, we wished to inquire into the extent to which the use of fire characterizes the archaeological sequence of GBY. Is this a unique phenomenon or a routine practice? The answer to this question may offer some insight into the hominins' technological capabilities with regard to the use, and particularly the control, of fire. In fact, we may never be able to ascertain whether early humans were "fire-makers" from the beginning. "It is very probable that the earliest Paleolithic fire-users were not fire-makers, but collected this precious commodity from natural conflagrations, and conserved it. Before man could utilize the accidental discovery that this or that action led to fire, he would have required some experience of handling it and this he could only have gained through having isolated and controlled fire of natural origin ..." (Oakley 1956:43, 1961:180–181). Thus, there is a significant difference between using fire and making fire at will. While the phantom hearths of GBY indeed provide evidence for the use of fire, it is difficult to determine with certainty whether this fire was "collected" by the Acheulian hominins from a natural source or whether they had the ability to set fire at will. However, as fire use is recorded in each of the occupational episodes at the site, it is unlikely that the hominins of GBY were compelled to collect it or rather to re-invent it over and over again. Rather, the fact that fire is repetitively used suggests that the knowledge of fire-making and the technological skills of the Acheulian hominins of GBY enabled them to set fire at will in diverse environmental settings (e.g., the storm beaches of Layers II-6 L-1–6 and the lake shores of Layers V-5 and V-6). The ability to make fire was most likely possessed by these Acheulian hominins throughout the long duration estimated for the entire stratigraphic sequence of the site (ca. 100 ka; Feibel 2001). The uninterrupted sequential use of fire at GBY is illustrated in Fig. 4.10. Concentrations of burned flint microartifacts are recorded in each of the eight occupational levels of Layer II-6. These archaeological levels are stratigraphically positioned one on top of the other and show no similarities in the locations of their phantom hearths (Fig. 4.10).

Fig. 4.10 The sequential occurrence of phantom hearths in Layer II-6 by level, from topmost level (II-6 L-1) to the lowermost level (II-6 L-7 upper occupational floor)

Finally, the results of this study have contributed to our understanding of the behavioral complexity of the hominins of GBY. Yet, on a more global perspective, the Acheulians of GBY are representatives of a fundamentally significant event in human evolution and dispersal. They are placed midway along the route out of Africa and into Eurasia, chronologically as well as geographically. Indeed, the Acheulian site of GBY displays the introduction of African lithic traditions to the Levantine Corridor, reflecting a wave of human migration out of Africa (see Section 1.3). The variety of evidence for the use of fire originating from African sites 1.5–1.0 Ma (see Section 1.1) may suggest that the use of fire at GBY reflects an additional African tradition. Moreover, the evidence from GBY may support the interpretation that views the use of fire as a triggering factor for human migration out of Africa and into Eurasia.

A date of ca. 1.8 Ma is generally suggested for the first dispersals out of Africa and into Eurasia and is also the base-line date of the so-called "long chronology" (see Dennell 2003 and references therein). Dennell (2003) conducted a thorough review of the indications for human presence outside East Africa during the Early Pleistocene and integrated evidence from North Africa, southern Asia, and Europe. This review suggested that during the Early Pleistocene the colonizing abilities of *Homo erectus* were very limited and that: "It was probably not until the Middle Pleistocene that hominids began habitually to utilize latitudes up to 45–50°N, which happen to include most of Europe" (Dennell 2003:435). The age and character of the earliest occupation of Europe are intensively discussed and debated (e.g., Carbonell et al. 1999, 2008; Dennell 2003; Roebroeks 2001; Roebroeks and van Kolfschoten 1995). Until recently, there was general agreement that the substantial settlement of Europe, in the form of continuous and isolated presence of hominins, occurred at ca. 0.5 Ma (Roebroeks and van Kolfschoten 1995; Dennell 2003). In this framework of "short chronology" (Roebroeks 2001) or "young Europe" (Carbonell et al. 1999), the dating or anthropogenic origin of sites claimed to precede this date was often challenged, while others were viewed as marginal, intermittent incursions. During the last decade, however, several discoveries (e.g., Atapuerca: Falguères et al. 1999; Parés and Pérez-González 1999; Fuente Nueva-3 and Barranco León: Oms et al. 2000; Ceprano: Manzi 2004) have supported the idea of a "long chronology" or "mature Europe". In this framework, early dispersals of Mode 1 technologies from Africa into Europe occurred as early as 1.0 Ma (Carbonell et al. 1996, 1999) and even reached northern parts of Europe (e.g., Pakefield, where however a Mediterranean climate is reconstructed: Parfitt et al. 2005).

Roebroeks (2001) stresses that: "The absence of Lower and early Middle Pleistocene sites north of the Pyrenees and Alps suggests that even if hominids were around the Mediterranean perimeter from the late Lower Pleistocene onwards, it required significant changes in their behaviour to take them north …" (Roebroeks 2001:454). Review of the early European evidence of fire use (see Chapter 1) indicates that the ability to control and maintain fire might well be the behavioral change that enabled the colonization of these northern parts of Europe, as suggested by Villa (1994, 2001): "Absence or non systematic use of fire may be one of the reasons why the settlement of Europe took a rather long time" (Villa 2001:4). Thus, the earlier incursions may have lacked the technological ability to make fire that would have enabled a more continuous settlement, like the one evident in the geographical and chronological extent of archaeological sites in Europe from 0.5 Ma onwards (see also Gowlett 2006).

The use of fire conferred various advantages on humans and probably enabled the colonization of new niches: "… with the possession of means of making fire at will man could freely leave his early circumscribed seat and successfully spread to other environments and eventually populate the earth" (Hough 1916:257).

Thus, having control over fire may have enabled hominin groups to migrate out of Africa and clearly played a major role in the colonization of Eurasia. These issues, as well as a variety of fire-related topics that transcend the realm of the hominins of GBY, are of great interest in the framework of human evolution and behavior. The hominins of GBY were part of a large-scale revolution that affected various aspects of human lives; the invention of fire had an impact on human diet, pyrotechnology, and socio-cultural behavior. In the case of GBY, where we are compelled to derive our interpretations from the limited archaeological data alone, we can only wonder about the technological, behavioral, and cultural consequences that these changes brought.

Appendices

Appendix 1: Burned and Unburned Flint Microartifacts

a–d: Layer V-5, burned flint microartifacts; e–h: Layer V-5, unburned flint microartifacts; i–l: Layer V-6, burned flint microartifacts; m–p: Layer V-6, unburned flint microartifacts

Appendix 2: Working Methodology for Arcmap GIS Files

In *ArcCatalog* (ESRI®ArcCatalog™9.3) *Italics – GIS terminology/function*

A

(1) Data preparation
Converting *Access* tables into geographic information for each database of lithic material
- (a) *Create feature class from x y table*
- (b) *Coordinate system*: Palestine 1923 Israel CS Grid

In *ArcMap* (ESRI®ArcMap™9.3) *Italics – GIS terminology/function*

A Defining extent of analyses

(1) Defining *General Environment*
- (a) *Coordinate system*: Palestine 1923 Israel CS Grid
- (b) Cluster tolerance: meters
- (c) Display: meters

(2) *Add data*
Inserting data layers into the ArcMap file
- (a) Insert the location map of excavation areas and trenches of GBY
- (b) Insert the feature class layers of microartifacts, macroartifacts, pebbles
- (c) Insert field map of living floor (when available)

(3) *Start editing*
Deleting items which occur outside the excavated area (i.e., registration errors)

(4) Defining the grid extent of the excavated units
- (a) Defining the rectangular extent of the archaeological material (i.e. X minimum/X maximum – Y minimum/Y maximum)
- (b) Creating a rectangular polygon grid which covers the extent of the archaeological material:
  Hawths Tools – Sampling Tools – Create Vector Grid
  Grid spacing: 0.5 × 0.5; lock 1:1 ratio
  Snap vector grid to major coordinate System

Projection definition: Palestine 1923 Israel CS Grid
The cells of the output "vector grid" match the excavated units (0.5 × 0.5 sub squares); each cell is a row in the table of "vector grid"

- (c) Defining the polygonal extent of the archaeological material (i.e., the excavated units)
  In ArcCatalog: *New – Shapefile – Polygon* – name of Shapefile: excavated units
  Add data: insert the excavated units shapefile
  Start editing: draw a polygon which borders the extent of the archeological material (i.e., includes all excavated units)
- (d) Clipping the vector grid to follow the excavated units
  Analysis tools – Extract – Clip

The area of the new "vector grid" includes all excavated units and is used for analyses which are based on frequencies per excavated unit (e.g., relative percentages per excavated unit)

(5) Defining the true extent of the excavated area
Draw a polygon which represents the extent of the excavated area according to the distribution of the archeological finds (available from the field maps of the living floors and from the distribution of coordinated pieces (i.e., items which were given a full XY coordinate in the field; these include FFT, CCT, bifacial tools and pebbles). The spatial extent of these items is the optimal representation of the boundaries of the excavated area of each archaeological layer
In ArcCatalog: *New – Shapefile – Polygon* – name of Shapefile: excavated area
Add data: insert the excavated area shapefile
Start editing: draw a polygon which borders the extent of coordinated archeological material
The outline of this polygon is illustrated in all distribution and density maps and is used as the extent for spatial analyses (e.g., kernel density maps – section D below)

B Creating distinct layers of microartifacts

(1) Separating the feature class of microartifacts into different layers by raw material and burning state
Selection – select by #attribute
#Raw material = flint/basalt/limestone
#Burning state = burned/unburned

(2) Creating five new layers according to raw material and burning state; each selection (B1) is followed by this stage:
Selection – create layer from selected features

The new layers are the point distribution of microartifacts: burned flint, unburned flint, total flint (burned + unburned flint), basalt, limestone

C Frequencies per excavated unit

(1) Counting the frequencies of flint microartifacts in each excavated unit

Hawths Tools – Analysis Tools – Count points in polygons
Input polygon layer: Vector Grid
Point Layer: burned flint (BR)/unburned flint (FL)/total flint (TF)/basalt (BS)/limestone (LM)
Output Field: PNTPOLYCBR/PNTPOLYCFL/PNTPOLYCTF/PNTPOLYCBS/PNTPOLYCLM

This function calculates the number of microartifacts (BR/FL/TF/BS/LM) in each cell of the grid and adds new fields into the table of vector grid where the counts of microartifacts are recorded

(2) Calculating the percentages of microartifacts in each excavated unit
(a) in the table view of vector grid *options – add field*: adds new empty fields to the table of vector grid
Fields names: %BR/%FL/%TF/%BS/%LM; fields properties – number of decimal places = 2
(b) In the table view of vector grid, right click an empty % field – *calculate values*:

For example, the formula to calculate the percentages of burned microartifacts in each excavated unit:

$$\%BR = \frac{(\text{PNTPOLYCBR} \times 100)}{\text{number of burned microartifacts in the layer}}$$

D Kernel density maps

(1) Creating kernel density maps of microartifacts
Spatial analyst – density – kernel
Point layer: burned flint microartifacts/unburned flint microartifacts
Cell size: 0.01 m
Search radius: 0.5 m
Area units: square meters
Search extent: excavated area

(2) Standardizing the density maps
Spatial analyst – raster calculator: regular kernel map/maximum value
Layer properties – symbology – classify: number of classes: 5; equal intervals

The values in the new kernel map are expressed on a scale from 0 to 1, with 1 representing the high-density kernel

(3) Creating interpolated density patterns of sub-square precision

In order to examine the validity of using the point plotted data for generating kernel density maps, we have tested the density patterns as illustrated based on the original recorded data (total counts per sub-square). Three different interpolation methods were applied (Appendix 3B):

(a) *Conversion tools – feature to raster*[1]
Input feature: vector grid
Field: PNTPOLYCBR
Output raster: vector_grid_ras_br
Cell size: 0.5

The values of the cells of the new raster correspond to the number of burned flint microartifacts in each excavated unit

(b) *Conversion tools – raster to points*
Input raster: vector_grid_ras_br
Field: value
Output point feature: br_subsq_points

The value of each point in the new point feature corresponds to the total counts of burned flint microartifacts of the sub square

(c1) *spatial analyst – Interpolation – IDW*
Input point feature: br_subsq_points
z value field: Grid Code (i.e., N of burned microartifacts)
Output raster: IDW_br_subsq
cell size: *0.01*
Search radius: 0.5
Extent: excavated area

(c2) *Spatial analyst – Interpolation – SPLINE*
Input point feature: br_subsq_points
z value field: Grid Code (i.e., N of burned microartifacts)
Output raster: SPLINE_br_subsq
Cell size: 0.01
Spline type: *tension*
Number of points: 12 (default)
Extent: excavated area

(c3) *Spatial analyst – Interpolation – KRIGING*
Input point feature: br_subsq_points
z value field: Grid Code (i.e., N of burned microartifacts)
Output raster: GRIGING_br_subsq
Kriging method: ordinary
Semivariogram model: spherical (default)
Cell size: 0.01
Search radius: 0.5
Extent: excavated area

E Homogeneity analysis: observed and expected burning

(1) Creating a raster layer of expected values of burned flint microartifacts
(a) In the table view of vector grid *options – add field* adds a new empty field to the table of vector grid;

[1] A raster is a spatial data set that defines space as an array of equally sized cells arranged in rows and columns. Each cell contains an attribute value and location coordinates. Unlike a vector structure, which stores coordinates explicitly, raster coordinates are contained in the ordering of the matrix.

field name: EXP_BR_N; right click on the empty field – *calculate values*:

$$EXP\_BR\_N = (PNTOPOLYTF) \times (\%OF\ BURNING / 100)$$

This uniformly calculates the percentage of burning of the particular analyzed layer out of the total number of flint microartifacts in each excavated unit; the new field thus records the expected *number* of burned flint microartifacts for each excavated unit in a case of a uniform distribution of the burning

(b) In the table view of vector grid *options – add field* adds a new empty field to the table of vector grid; field name: EXP_BR_%; right click on the empty field – *calculate values*:

$$EXP\_BR\_\% = \frac{EXP\_BR\_N \times 100}{\text{number of burned flint microartifacts in layer}}$$

This calculates the relative percentage of the expected number of burned flint microartifacts from the total number of burned flint microartifacts of the layer within each excavated unit; this is the expected relative percentage of burned flint microartifacts in a case of uniform distribution of the burning

(c) *Spatial analyst – convert – features to raster*
Input feature: vector grid
Field: EXP_BR_%
Cell size: 0.5
Output raster: EXPECTED_BR_%

The new raster layer depicts the expected percentages of burned flint microartifacts within each excavated unit

(2) Creating a raster layer of observed values of burned flint microartifacts
(a) *Spatial analyst – convert – features to raster*
Input feature: vector grid
Field: %BR (see C2 above)
Cell size: 0.5
Output raster: OBSERVED_BR_%

The new raster layer depicts the observed percentages of burned flint microartifacts within each excavated unit

(3) Calculating the deviation between the observed and expected percentages of burned flint microartifacts in each excavated unit
Spatial analyst – raster calculator:

OBSERVED_BR_% – EXPECTED_BR_%

This subtracts the expected percentage of burned flint microartifacts raster from the observed percentage of burned flint microartifacts raster; in the new output raster, the value of each cell is the deviation between the observed and expected percentage of burning; cells of positive values are excavated units in which the observed percentage of burning exceeds the expected one in a case of uniform distribution of the burned flint microartifacts

F statistical tests

(1) Creating a raster layer of the chi square test values for the burned flint microartifacts in each excavated unit
(a) Creating a raster layer of observed numbers of burned flint microartifacts per excavated unit *spatial analyst – convert – features to raster*
Input feature: vector grid
Field: PNTPOLYCBR (C1 above)
Cell size: 0.5
Output raster: OBSERVED_BR_N
(b) Calculating chi square values

This calculation is creating a new raster layer based on the values recorded in the raster layers of OBSERVED_BR_N (F1 above) and EXP_BR_N (E1 above)
Spatial analyst – raster calculator:

$$\chi^2 = \sum_i \frac{(OBS_i - EXP_i)^2}{EXP_i}$$

The absolute (i.e., Σ) chi square test value of a particular archaeological layer is the summary of χ^2 values of all excavated units (i = number of excavated units).

(2) Creating a raster layer with the standardized residuals (SR; the signed square root of each category's contribution to the χ^2) values for the burned flint microartifacts in each excavated unit
Spatial analyst – raster calculator:

$$SR = \frac{OBS_i - EXP_i}{\sqrt{EXP_i}}$$

G Random plotting

(1) Creating three sets of randomly selected points. The number of points (N) is equivalent to the number of burned flint microartifacts recorded in the analyzed layer:
(a) *Hawths tools – sampling – create random selection*
Layer to select features in: total flint (see B2 above)
This number of features: N; equivalent to the number of burned flint microartifacts recorded in the analyzed layer
Selection – create layer from selected features
This procedure is sequentially repeated 3 times
Name of layer: RANDOM 1/RANDOM 2/RANDOM 3

(2) Creating kernel density maps of randomly selected microartifacts
Spatial analyst – density – kernel
Point layer: RANDOM 1/RANDOM 2/ RANDOM 3
Cell size: 0.01 m
Search radius: 0.5 m
Area units: square meters
Search extent: excavated area

(3) Standardizing the random density maps
Spatial analyst – raster calculator: kernel map/maximum value
Layer properties – symbology – classify: number of classes: 5; equal intervals

The values in the new kernel map are expressed on a scale from 0 to 1, with 1 representing the high-density kernel

Appendix 3A: The use of different search radii and cell sizes in kernel density maps

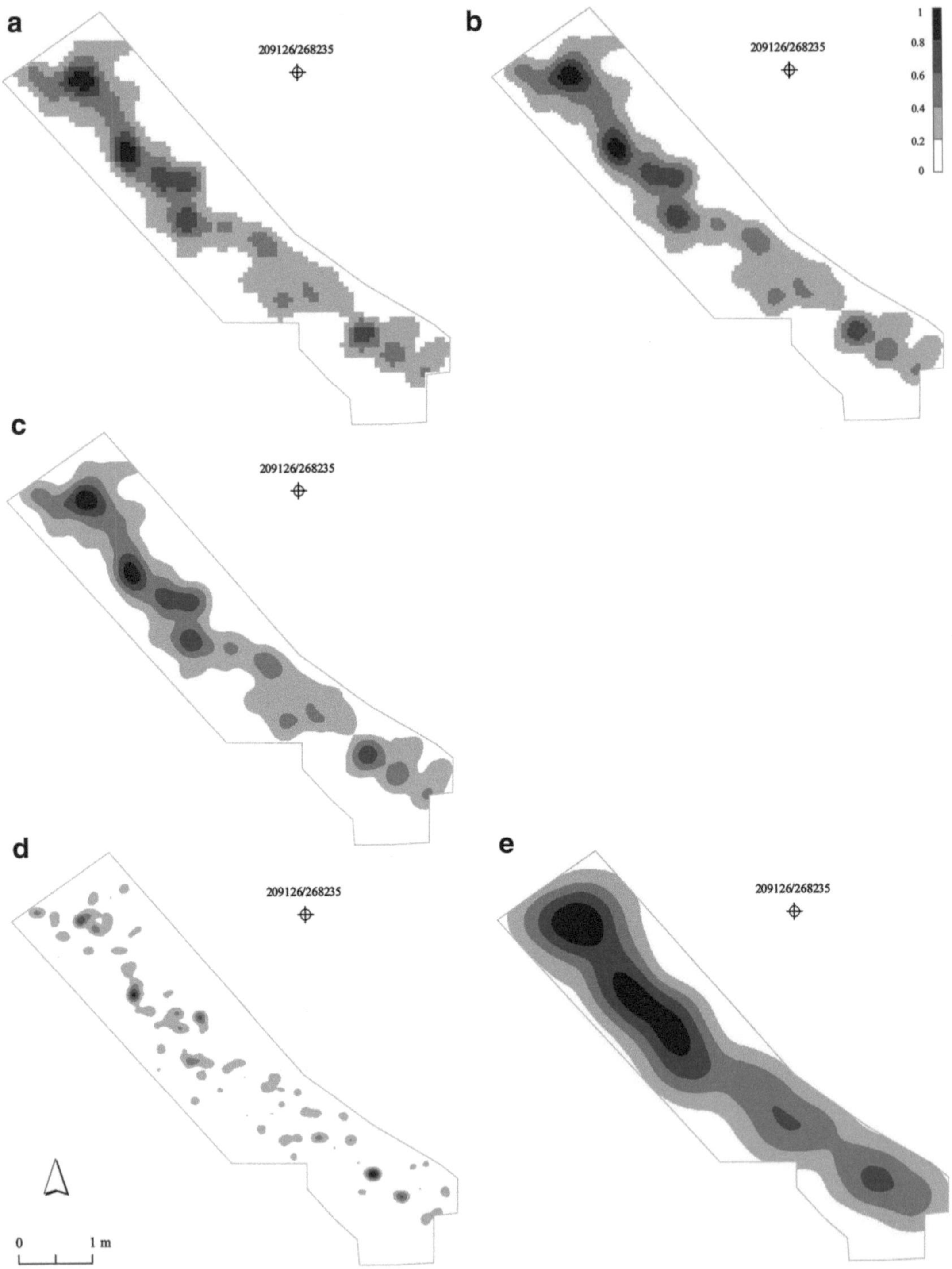

Kernel density maps produced for the assemblage of burned flint microartifacts of Layer II-6 L-1 (N = 754) with different search radii and cell sizes (units in meters); (a) cell size = 0.1, search radius = 0.5; (b) cell size = 0.5, search radius = 0.5; (c) The cell size and search radius used in this study: cell size = 0.01, search radius=0.5; (d) cell size = 0.01, search radius = 0.2; and (e) cell size = 0.01, search radius = 1

Appendix 3B: Density maps based on data interpolation of sub-square precision

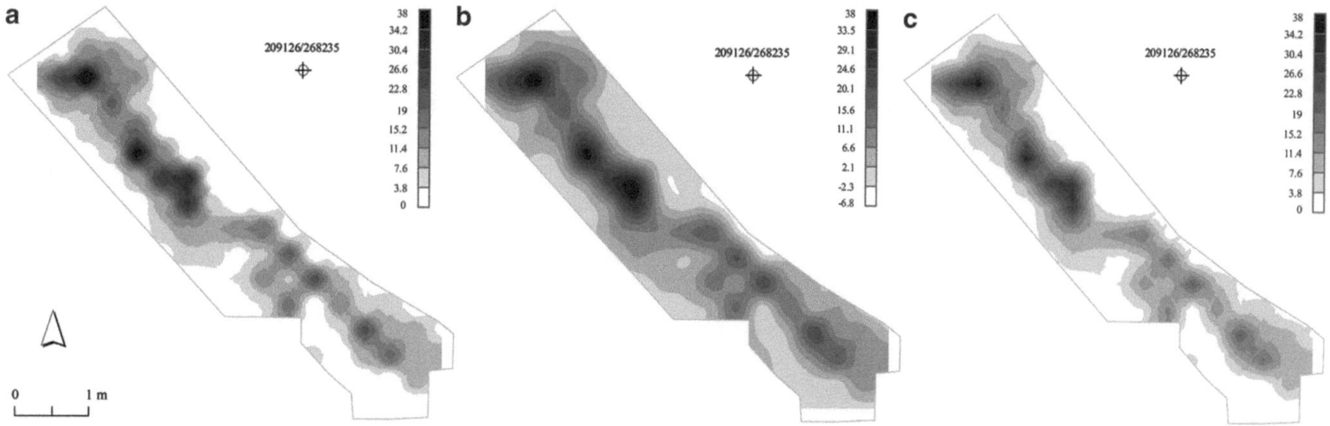

Density maps in which the density patterns are depicted through data interpolation of sub-square precision (total counts per sub-square) were produced in order to test the validity of using kernel density maps (which use point plotted data). This examination was carried out on the assemblage of burned flint microartifacts from Layer II-6 L-1 (N = 754). (a) IDW interpolation; (b) Spline interpolation; and (c) Kriging interpolation

Appendix 4: Probability of a Natural Wildfire at GBY

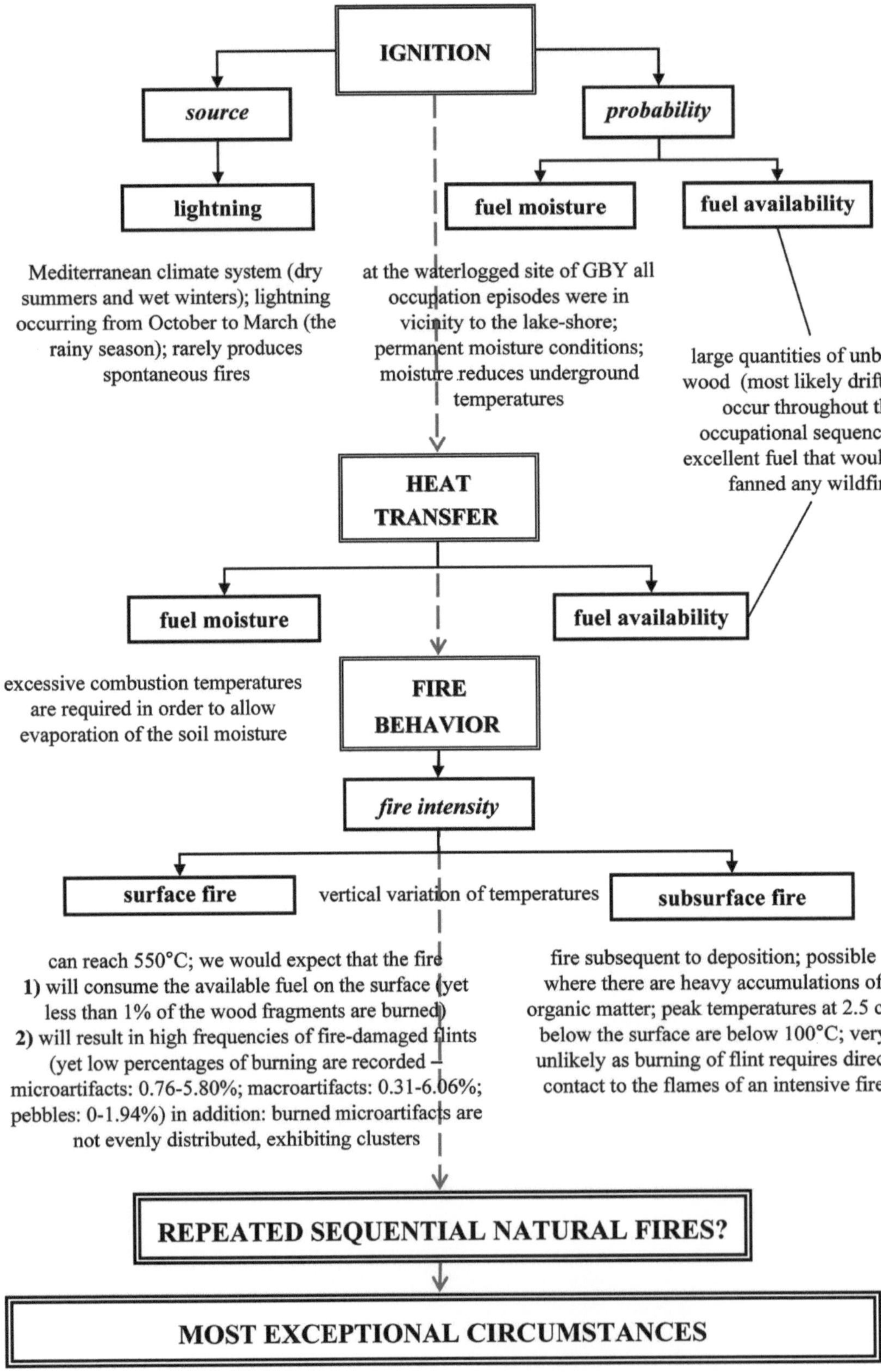

References

Aiello LC, Wheeler P. The expensive-tissue hypothesis. Curr Anthropol. 1995;36:199–221.

Albert RM, Lavi O, Estroff L, Weiner S, Tsatskin A, Ronen A, et al. Mode of occupation of Tabun Cave, Mt. Carmel, Israel during the Mousterian period: a study of the sediments and phytoliths. J Archaeol Sci. 1999;26:1249–60.

Albert RM, Weiner S, Bar-Yosef O, Meignen L. Phytoliths in the Middle Palaeolithic deposits of Kebara Cave, Mt. Carmel, Israel: study of the plant materials used for fuel and other purposes. J Archaeol Sci. 2000;27:931–47.

Albert RM, Bar-Yosef O, Meignen L, Weiner S. Quantitative phytolith study of hearths from the Natufian and Middle Palaeolithic levels of Hayonim Cave (Galilee, Israel). J Archaeol Sci. 2003;30:461–80.

Alperson-Afil N, Hovers E. Differential use of space in the Neandertal site of Amud Cave, Israel. Eurasian Prehist. 2005;3(1):3–22.

Alperson-Afil N, Richter D, Goren-Inbar N. Phantom hearths and the use of fire at Gesher Benot Ya'aqov, Israel. PaleoAnthropol. 2007;1:1–15. http://www.paleoanthro.org/journal/content/PA20070001.pdf.

Ambret P. Utilisation préhistorique de la technique minière d'abattage au feu dans le district cuprifère de Cabrières (Hérault). Comptes Rendus Palevol. 2002;1:711–6.

Ashkenazi S, Motro U, Goren-Inbar N, Biton R, Rabinovich R. New morphometric parameters for assessment of body size in the fossil freshwater crab assemblage from the Acheulian site of Gesher Benot Ya'aqov, Israel. J Archaeol Sci. 2005;32:675–89.

Ashton N, Cook J, Lewis SG, Rose J. High Lodge. Excavations by G. Sieveking, 1962–1968 and J Cook, 1988. London: British Museum Press; 1992.

Ashton N, Lewis SL, Parfitt S (editors). Excavations at the Lower Palaeolithic Site at East Farm, Barnham, Suffolk, 1989–1994. London: British Museum Occasional Paper 125; 1998.

Austin LA, Bergman CA, Roberts MB, Wilhelmsen KH. Archaeology of excavated areas. In: Roberts MB, Parfitt SA, editors. Boxgrove: a Middle Pleistocene hominid site at Eartham Quarry, Boxgrove, West Sussex. London: English Heritage; 1999. p. 312–77.

Bachelard G. The psychoanalysis of fire. London: Routledge & Kegan Paul; 1938 (1964).

Barbetti M. Traces of fire in the archaeological record before one million years ago? J Hum Evol. 1986;15:771–81.

Barkai R, Gopher A, Lauritzen SE, Frumkin A. Uranium series dates from Qesem Cave, Israel, and the end of the Lower Palaeolithic. Nature. 2003;423:977–9.

Bar-Yosef O, Goren-Inbar N. The Lithic assemblages of 'Ubeidiya: a Lower Palaeolithic site in the Jordan Valley. Qedem Volume 34, Monographs of the Institute of Archaeology, Jerusalem: Hebrew University; 1993.

Beaumont PB, Vogel JC. On a timescale for the past million years of human history in central South Africa. S Afr J Sci. 2006; 102:217–28.

Belitzky S, Goren-Inbar N, Werker E. A Middle Pleistocene wooden plank with man-made polish. J Hum Evol. 1991;20:349–53.

Bellomo RV. Early Pleistocene fire technology in Northern Kenya. In: Childs TS, editor. Society culture and technology in Africa. Philadelphia: University of Pennsylvania; 1994a. p. 17–28.

Bellomo RV. Methods of determining early hominid behavioral activities associated with the controlled use of fire at FxJj 20 Main, Koobi Fora, Kenya. J Hum Evol. 1994b;27:173–95.

Bellomo RV, Harris JWK. Preliminary report of actualistic studies of fire within Virunga National Park, Zaire: towards an understanding of archaeological occurrences. In: Boaz NT, editor. Evolution of environments and Hominidae in the African Western Rift Valley. Martinsville: Virginia Museum of Natural History; 1990. p. 317–38.

Bennet JL. Thermal alternation of buried bone. J Archaeol Sci. 1999; 26:1–8.

Binford LR. Dimensional analysis of behavior and site structure: learning from an Eskimo hunting stand. Am Antiquity. 1978; 43:330–61.

Binford LR. In pursuit of the past: decoding the archaeological record. London: Thames & Hudson; 1983.

Binford LR. Butchering, sharing, and the archaeological record. J Anthropol Archaeol. 1984;3:235–57.

Binford LR. Hearth and home: the spatial analysis of ethnographically documented rock shelter occupations as a template for distinguishing between human and hominid use of sheltered space. XIII U.I.S.P.P Congress, 8–14 September 1996, Forly. Congress Proceedings; 1998. p. 229–39.

Bischoff JL, Ikeya M, Budinger FE. A TL/ESR study of the hearth feature at the Calico archaeological site, California. Am Antiquity. 1984;49:764–74.

Blainey G. Triumph of the nomads: a history of ancient Australia. Melbourne: Macmillan; 1975.

Blasco López R. Estrategias de Subsistencia de los Hominidos del Nivel XII de la Cova del Bolomor (la Valldigna, Valencia). Unpublished Ph.D. thesis. Tarragona: Universitat Rovira i Virgili; 2006.

Blondel J, Aronson J. Biology and wildlife of the Mediterranean region. Oxford: Oxford University Press; 1999.

Boback SM, Cox CL, Ott BD, Carmody R, Wrangham RW, Secor SM. Cooking and grinding reduces the cost of meat digestion. Comp Biochem Physiol Part A. 2007;148:651–56.

Boëda E, Connan J, Dessort D, Muhesen S, Mercier N, Valladas H, et al. Bituman as a hafting material on Middle Paleolithic artifacts. Nature. 1996;380:336–38 Bonhomme T, Stanley J. Magnetic mapping of prehistoric aboriginal fireplaces at Bunda Lake, Belarabon Station, Western New South Wales. Aus Archaeol. 1985;21:63–73.

Boyd M. Identification of anthropogenic burning in the paleoecological record of the Northern Prairies: a new approach. Ann Assoc Am Geogr. 2002;92:471–87.

Brain CK, Sillen A. Evidence from the Swartkrans Cave for the earliest use of fire. Nature. 1988;336:464–6.

Brande S, Saragusti I. A morphometric model and landmark analysis of Acheulian hand axes from Northern Israel. In: Marcus LF, Corti M,

Loy A, Naylor GJP, Slice DE, editors. Advances in morphometrics. New York: Plenum; 1996. p. 423–35.
Breuil H. Le feu et l'industrie de pierre et d'os dans le gisement du "Sinanthropus" à Chou Kou Tien. L'Anthropologie. 1932;42:1–17.
Brewer S. The Chimps of Mt. Asserik. New York: Alfred A Knopf; 1978.
Briles CE, Whitlockb C, Bartlein PJ. Postglacial vegetation, fire, and climate history of the Siskiyou Mountains, Oregon, USA. Quatern Res. 2005;64:44–56.
Brooks AS, Yellen JE. The preservation of activity areas in the archaeological record: ethnoarchaeological and archaeological work in Northwest Ngamiland, Botswana. In: Kent S, editor. Method and theory for activity area research: an ethnoarchaeological approach. New York: Columbia University Press; 1987. p. 63–106.
Bullinger J, Leesch D, Plumettaz N. Le Site Magdalénien de Monruz: Premiers Éléments pour l'Analyse d'un Habitat de Plein Air. Neuchâtel: Service et Musée Cantonal d'Archéologie; 2006.
Bunn HT, Kroll EM. Systematic butchery by Plio/Pleistocene hominids at Olduvai Gorge, Tanzania. Curr Anthropol. 1986;7:431–52.
Callow P, Walton D, Shell CA.The use of fire at La Cotte de St. Brelade. In: Callow P, Cornford JM, editors. La Cotte de St. Brelade 1961–1978. Excavations by CBM McBurney. Norwich: Geo Books; 1986. p. 193–5.
Canti MG, Linford N. The effects of fire on archaeological soils and sediments: temperture and colour relationships. Proc Prehist Soc. 2000;66:385–95.
Carbonell E, Mosquera M, Rodríguez XP, Sala R. The first human settlement of Europe. J Anthropol Res. 1996;52:107–14.
Carbonell E, Mosquera M, Rodríguez XP, Sala R, van der Made J. Out of Africa: the dispersal of the earliest technical systems reconsidered. J Anthropol Archaeol. 1999;18:119–36.
Carbonell E, Mosquera M, Rodríguez XP, Bermúdez de Castro JM, Burjachs F, Rosell J, et al. Eurasian gates: the earliest human dispersals. J Anthropol Res. 2008;64:195–28.
Cessford C. Microartifactual floor patterning: the case at Catalhoyuk; 2003. http://www.shef.ac.uk/assem/issue7/cessford.html 7, 1–13.
Clark JD. Acheulian occupation sites in the Middle East and Africa: a study in cultural variability. In: Clark JD, Howell FC, editors. Recent studies in paleoanthropology. Wisconsin: American Anthropological Association; 1966. p. 202–29.
Clark JD. The Middle Acheulian occupation site at Latamne Northern Syria (second paper): further excavations (1965) – general results, definition and interpretation. Quaternaria. 1968;10:1–71.
Clark JD. Kalambo Falls prehistoric site. Volume I. Cambridge: Cambridge University Press; 1969.
Clark JD. Modified, used and other wood and plant specimens from Bwalya Industry Acheulean horizons. In: Clark JD, editor. Kalambo Falls prehistoric site Volume III – the earlier cultures: Middle and Earlier Stone Age. Cambridge: Cambridge University Press; 2001. p. 481–91.
Clark JD, Harris JWK. Fire and its roles in early hominid lifeways. Afr Archaeol Rev. 1985;3:3–27.
Clark JD, Asfaw B, Assefa G, Harris JWK, Kurashina H, Walter RC, et al. Palaeoanthropological discoveries in the Middle Awash Valley, Ethiopia. Nature. 1984;307:423–8.
Clark JE. Flintknapping and debitage disposal among the Lacandon Maya of Chiapas, Mexico. In: Staski E, Sutro LD, editors. The ethnoarchaeology of refuse disposal. Arizona: Arizona State University; 1991. p. 63–78.
Cohen-Ofri I, Weiner L, Boaretto E, Mintz G, Weiner S. Modern and fossil charcoal: aspects of structure and diagenesis. J Archaeol Sci. 2006;33:428–39.
Copeland L. The 1958 excavation of Abri Zumoffen in retrospect and the extension of 1963. In: Roe DA, editor. Adlun in the Stone Age: the excavations of D. A. E. Garrod in the Lebanon, 1958–1963. BAR International Series. Volume 159. Oxford: Archaeopress; 1983. p. 75–86.
Copeland L. Yabrudian and related industries: the state of research in 1996. In: Ronen A, Weinstein-Evron M, editors. Toward modern humans – the Yabrudian and Micoquian 400–50 k-years ago. Proceedings of a congress held at the University of Haifa 3–9 November 1996. BAR International Series. Volume 850. Oxford: Archaeopress; 2000. p. 97–117.
Courty M-A. Analyse microscopique des sédiments du remplissage de la grotte Vaufrey (Dordogne). In: Rigaud J-P, editors. La Grotte Vaufrey. Mémoires de la Société Préhistorique Française 19; 1988. p. 183–298.
Crabtree DE, Butler BR. Notes on experiments in flint knapping I: heat treatments of silica minerals. Tebiwa. 1964;7(1):1–6.
Day GM. The Indians as an ecological factor in the Northeastern forest. Ecology. 1953;34:329–46.
de Lumley H. Une cabane acheuléenne dans la Grotte du Lazaret (Nice). Mémoires de la Société Préhistorique Française. 1969;7:1–236.
de Lumley H. Il y a 400 000 ans: la domestication du feu, un formidable moteur d'hominisation. Comptes Rendus Palevol. 2006;5:149–54.
DeBano LF, Neary DG, Ffolliott PF. Fire's effects on ecosystems. New York: Wiley; 1998.
DeBoer WR. The archaeological record as preserved death assemblage. In: Moore JA, Keene AS, editors. Archaeological hammers and. theories. New York: Academic; 1983. p. 19–36.
Delagnes A, Roche H. Late Pliocene hominid knapping skills: The case of Lokalalei 2C, West Turkana, Kenya. J Hum Evol. 2005;48:435–72.
Delcourt PA, Delcourt HR, Ison CR, Sharp WE, Gremillion KJ. Prehistoric human use of fire, the eastern agricultural complex, and. Appalachian oak-chestnut forests: paleoecology of Cliff Palace Pond, Kentucky. Am Antiquity. 1998;63:263–78.
Delson E. Chronology of South African Australopith site units. In: Grine FE, editor. Evolutionary history of the "Robust" Australopitheciness. New York: Aldine; 1988. p. 317–24.
Dennell RW. Comments on "Hominid use of fire in the Lower and Middle Pleistocene: review of the evidence" (James, S.R). Curr Anthropol. 1989;30:11–3.
Dennell RW. Dispersal and colonisation, long and short chronologies: how continuous is the Early Pleistocene record for hominids outside East Africa? J Hum Evol. 2003;45:421–40.
Díez JC, Fernández-Jalvo Y, Rosell J, Cáceres I. Zooarchaeology and taphonomy of Aurora stratum (Gran Dolina, Sierra de Atapuerca, Spain). J Hum Evol. 1999;37:623–53.
Domínguez-Rodrigo M, Pickering TR, Semaw S, Rogers MJ. Cutmarked bones from Pliocene archaeological sites at Gona, Afar, Ethiopia: implications for the function of the world's oldest stone tools. J Hum Evol. 2005;48:109–21.
Dunnell RC, Stein JK. Theoretical issues in the interpretation of microartifacts. Geoarchaeology. 1989;4:31–42.
Elbaum R, Weiner S, Albert RM, Elbaum M. Detection of burning of plant materials in the archaeological record by changes in the refractive indices of siliceous phytoliths. J Archaeol Sci. 2003;30:217–26.
Fairbairn AS, Hope GS, Summerhayes GR. Pleistocene occupation of New Guinea's highland and subalpine environments. World Archaeol. 2006;38:371–86.
Falguères C, Yokoyama Y, Quaegebeur JP. Datations de sédiments quaternaires par la méthode de résonance de spin éléctronique (ESR). L'Anthropologie. 1988;92:723–26.
Falguères C, Bahain J-J, Yokoyama Y, Arsuaga JL, Bermudez de Castro JM, Carbonell E, et al. Earliest humans in Europe: the age of TD6 Gran Dolina, Atapuerca, Spain. J Hum Evol. 1999;37:343–52.
Falguères C, Yokoyama Y, Shen G, Bischoff JL, Ku T-L, de Lumley H. New U-series dates at the Caune de l'Arago, France. J Archaeol Sci. 2004;31:941–52.
Feibel CS. The Acheulian site of Gesher Benot Ya'aqov. Volume IV. Sediments and paleoenvironment; in preparation.
Feibel CS. Archaeological sediments in lake margin environments. In: Stein JK, Farrand WR, editors. Sediments in archaeological contexts. Salt Lake City: University of Utah Press; 2001. p. 127–48.

Feibel CS. Quaternary lake margins of the Levant Rift Valley. In: Goren-Inbar N, Speth JD, editors. Human paleoecology in the Levantine Corridor. Oxford: Oxbow Books; 2004. p. 21–36.

Fernández Peris J. Cova del Bolomor (La Valldigna, Valencia). Un registro paleoclimático y arqueológico en un medio Kárstico. Sociedad Española De Espeleología Y Ciencias Del Karst Boletín 4 Sedeck; 2003. p. 34–47.

Fernández-Jalvo Y, Díez JC, Rosell J, Cáceres I. Human cannibalism in the Early Pleistocene of Europe (Gran Dolina, Sierra de Atapuerca, Spain). J Hum Evol. 1999;37:591–622.

Fernández-Peris J. La Cova del Bolomor. Valencia: Diputación provincial de Valencia; 2007.

Figueiral I, Mosbrugger V. A review of charcoal analysis as a tool for assessing Quaternary and Tertiary environments: achievements and limits. Palaeogeogr Palaeoclimatol Palaeoecol. 2000;164: 397–407.

Fladmark KR. Microdebitage analysis: initial considerations. J Archaeol Sci. 1982;9:205–20.

Forbes RJ. Studies in ancient technology. Volume VII. Leiden: E. J. Brill; 1963.

Frazer JG. Myths of the origin of fire. London: MacMillan; 1930.

Freeman LG. Acheulean sites and stratigraphy in Iberia and the Magreb. In: Butzer KW, Isaac GL editors. After the Australopithecines. Stratigraphy, ecology, and culture change in the Middle Pleistocene. Hague: Mouton; 1975. p. 661–743.

Galanidou N. Patterns in caves: foragers, horticulturists, and the use of space. J Anthropol Archaeol. 2000;19:243–75.

Galanidou N. "Home is where the hearth is": the spatial organization of the Upper Palaeolithic rockshelter occupations at Klithi and Kastritsa in Northwest Greece. BAR International Series. Volume 687. Oxford: Archaeopress; 1997.

Garrod DAE, Bate DMA. The Stone Age of Mount Carmel. London: Clarendon; 1937.

Garrod D, Kirkbride D. Excavations of the Abri Zumoffen, a Rock-Shelter near Adlun, South Lebanon, 1958. Beyrouth: Bulletin du Musée de Beyrouth Volume XVI; 1961.

Geigel EM. On the circumstances surrounding the preservation and analysis of very old DNA. Archaeometry. 2002;44:337–42.

Geigel EM, Baumer U, Koller J. New approaches to study the preservation of biopolymers in fossil bones. Environ Chem Lett. 2004; 2:45–8.

Geraads D, Tchernov E. Femurs humains du Pleistocène Moyen de Gesher Benot Ya'aqov (Israël). L'Anthropologie. 1983;87:138–41.

Gilead I. A Middle Paleolithic open-air site near Tell Far'a, Western Negev: preliminary report. Isr Expl J. 1980;30:52–62.

Gilead I. Too many notes? virtual recording of artifacts provenance. In: Niccolucci F, editor. Virtual archaeology: proceedings of the VAST euroconference, Arezzo. BAR International Series. Volume 1075. Oxford: Archaeopress; 24–25 November 2000. p. 41–43.

Gilead I, Grigson C. Far'a II: a Middle Paleolithic open-air site in the Northern Negev, Israel. Proc Prehist Soc. 1984;50:71–97.

Goldberg P, Weiner S, Bar-Yosef O, Xu Q, Liu J. Site formation processes at Zhoukoudian, China. J Hum Evol. 2001;41:483–530.

Goren-Inbar N. The Gesher Benot Ya'aqov Acheulian site – the environment and behavior of Early/Middle Pleistocene settlers in the Hula Valley. In: Derevianko AP, Nokhrina TI, editors. Arkheologiya i Paleoekologiya Evrasii [Archaeology and Paleoecology of Eurasia]. Novosibirsk: Institute of Archaeology and Ethnography SB RAS Press; 2004. p. 267–78.

Goren-Inbar N, Belitzky S. Structural position of the Pleistocene Gesher Benot Ya'aqov site in the Dead Sea Rift Zone. Quatern Res. 1989;31:371–76.

Goren-Inbar N, Saragusti I. An Acheulian biface assemblage from the site of Gesher Benot Ya'aqov, Israel: indications of African affinities. J Field Archaeol. 1996;23:15–30.

Goren-Inbar N, Sharon G. Invisible handaxes and visible Acheulian biface technology at Gesher Benot Ya'aqov, Israel. In: Goren-Inbar N, Sharon G, editors. Axe age: Acheulian tool-making from quarry to discard. Oxford: Oxbow Books; 2006. p. 111–35.

Goren-Inbar N, Belitzky S, Verosub K, Werker E, Kislev M, Heimann A, et al. New discoveries at the Middle Pleistocene Gesher Benot Ya'aqov Acheulian site. Quatern Res. 1992;38:117–28.

Goren-Inbar N, Lister A, Werker E, Chech M. A butchered elephant skull and associated artifacts from the Acheulian site of Gesher Benot Ya'aqov, Israel. Paléorient. 1994;20(1):99–12.

Goren-Inbar N, Feibel CS, Verosub KL, Melamed Y, Kislev ME, Tchernov E, et al. Pleistocene milestones on the out-of-Africa corridor at Gesher Benot Ya'aqov, Israel. Science. 2000;289:944–47.

Goren-Inbar N, Sharon G, Melamed Y, Kislev M. Nuts, nut cracking, and pitted stones at Gesher Benot Ya'aqov, Israel. PNAS. 2002a; 99:2455–60.

Goren-Inbar N, Werker E, Feibel CS. The Acheulian site of Gesher Benot Ya'aqov. Volume I. The wood assemblage. Oxford: Oxbow Books; 2002b.

Goren-Inbar N, Alperson N, Kislev ME, Simchoni O, Melamed Y, Ben-Nun A, et al. Evidence of hominin control of fire at Gesher Benot Ya'aqov, Israel. Science. 2004;304:725–27.

Goren-Inbar N, Sharon G, Alperson-Afil N, Laschiver I. The Acheulian massive scrapers of Gesher Benot Ya'aqov – a product of the bifaces' chaîne opératoire. J Hum Evol. 2008;55:702–12.

Goring-Morris AN. Prehistoric Investigations in the Western Negev. Part 1: The Shunera Dunes, Givat Hayil, Nahal Sekher and Hamifgash. Qedem. Jerusalem: Monographs of the Institute of Archaeology, Hebrew University; in preparation.

Goring-Morris AN. Trends in spatial organization of terminal Pleistocene hunter-gatherer occupations as viewed from the Negev. and Sinai. Paléorient. 1988;14(2):231–44.

Goudsblom J. The human monopoly on the use of fire: its origins and conditions. Hum Evol. 1986;1:517–23.

Goudsblom J. Fire and civilization. London: Penguin Books; 1992.

Gowlett JAJ. The early settlement of northern Europe: fire history in the context of climate change and the social brain. Comptes Rendus Palevol. 2006;5:299–310.

Gowlett JAJ, Harris JWK, Walton D, Wood BA. Early archaeological sites, hominid remains and traces of fire from Chesowanja, Kenya. Nature. 1981;294:125–29.

Gowlett JAJ, Hallos J, Hounsell S, Brant V, Debenham NC. Beeches Pit: Archaeology, assemblage dynamics and early fire history of a Middle Pleistocene site in East Anglia, UK. Eurasian Prehist. 2005;3(2):3–38.

Grayson DK. The archaeological record of human impacts on animal populations. J World Prehist. 2001;15:1–68.

Green HJM. An analysis of archaeological rubbish deposits. Archaeol News Lett. 1961;7(4):91–3.

Haberle SG, Hope GF, Van Der Kaars S. Biomass burning in Indonesia and Papua New Guinea: natural and human induced fire events in the fossil record. Palaeogeogr Palaeoclimatol Palaeoecol. 2001; 171:259–68.

Haberman SJ. The analysis of residuals in cross-classified tables. Biometrics. 1973;29:205–20.

Hanson M, Cain CR. Examining histology to identify burned bone. J Archaeol Sci. 2007;34:1902–13.

Harrison HS. Fire-making, fuel, and lighting. In: Singer C, Holmyard EJ, Hall AR, editors. A history of technology. Oxford: Clarendon; 1954. p. 216–37.

Hayden B, Cannon A. Where the garbage goes: refuse disposal in the Maya highlands. J Anthropol Archaeol. 1983;2:117–63.

Hedgcock FT, Marshall H, Berger GW, Smith PEL. Burnt stones from ancient fires. In: Farquhar RM, Hancock RGV, Pavlish LA, editors. Proceedings of the 26th international archaeometry symposium. Toronto: University of Toronto; 1988. p. 48–52.

Hietala H. Boker Tachtit: spatial distributions. In: Marks AE, editor. Prehistory and paleoenvironments in the Central Negev, Israel. Dallas: Southern Methodist University; 1983. p. 191–216.

Hillman GC, Robins GV, Oduwole D, Sales KD, McNeil DAC. Determination of thermal histories of archeological cereal grains with electron spin resonance spectroscopy. Science. 1983;222: 1235–6.

Hough W. The distribution of man in relation to the invention of fire-making methods. Am Anthropol. 1916;18:257–63.

Howell FC. Observations on the earlier phases of the European Lower Paleolithic. Am Anthropol. 1966;68:88–201.

Hull KL. Identification of cultural site formation processes through microdebitage analysis. Am Antiquity. 1987;52:772–83.

Huxtable J. The thermoluminescence dates. In: Callow P, Cornford JM, editors. La Cotte de St. Brelade 1961–1978. Excavations by CBM McBurney. Norwich: Geo Books; 1986. p. 145–9.

Hyodo M, Nakaya H, Urabe A, Saegusa H, Shunrong X, Jiyun Y, et al. Paleomagnetic dates of hominid remains from Yuanmou, China, and other Asian sites. J Hum Evol. 2002;43:27–41.

Inizan ML. Outil lithiques Caspiens Ocrés. L'Anthropologie. 1976; 80:39–63.

Isaac GL. Olorgesailie: archeological studies of a Middle Pleistocene Lake Basin in Kenya. Chicago, IL: The University of Chicago Press; 1977.

Isaac GL. The food-sharing behavior of protohuman hominids. Sci Am. 1978;238:90–108.

Isaac GL. Early hominids and fire at Chesowanja, Kenya. Nature. 1982;296:870.

Isaac GL. The archaeology of human origins: studies of the Lower Pleistocene in East Africa: 1971–1981. In: Wendorf F, Close AE, editors. Advances in world archaeology, vol. 3. New York: Academic; 1984. p. 1–87.

Isaac GL. Bones in contention: competing explanations for the juxtaposition of Early Pleistocene artefacts and faunal remains. In: Isaac B, editor. The archaeology of human origins – papers by Glynn Isaac. Cambridge: Cambridge University Press; 1983 (1989). p. 325–35.

Isaac GL, Harris JWK. Archaeology. In: Leakey MD, Leakey REF, editors. Koobi Fora research project, vol. 1. Oxford: Clarendon; 1978. p. 64–85.

James SR. Hominid use of fire in the Lower and Middle Pleistocene. Curr Anthropol. 1989;30:1–26.

Jordanova N, Petrovsky E, Kovacheva M, Jordanova D. Factors determining magnetic enhancement of burnt clay from archaeological sites. J Archaeol Sci. 2001;28:1137–48.

Julig PJ, Long DGF, Schroeder B, Rink WJ, Richter D, Schwarcz HP. Geoarchaeology and new research at Jerf al-Ajla Cave, Syria. Geoarchaeology. 1999;14:821–48.

Kaplan H, Hill K, Lancaster J, Hurtado M. A theory of human life history evolution: diet, intelligence, and longevity. Evol Anthropol. 2000;9:156–85.

Karkanas P, Rigaud J-P, Simek JF, Albert RM, Weiner S. Ash, bones and guano: a study of the minerals and phytoliths in the sediments of Grotte XVI, Dordogne, France. J Archaeol Sci. 2002;29:721–32.

Karkanas P, Shahack-Gross R, Ayalon A, Bar-Matthews M, Barkai R, Frumkin A, et al. Evidence for habitual use of fire at the end of the Lower Paleolithic: site-formation processes at Qesem Cave, Israel. J Hum Evol. 2007;53:197–212.

Kirkbride D. The soundings in Bezez Cave. In: Roe DA, editor. Adlun in the Stone Age. Excavations of DAE Garrod in the Lebanon, 1958–1963. BAR International Series. Volume 159. Oxford: Archaeopress; 1983. p. 23–67.

Kleindienst MR. Components of the East African Acheulian assemblages: an analytic approach. In: Mortelmans G, editor. Actes du IVème Congrès Panafricain de Préhistoire et de l'Etude du Quaternaire. Tervuren: Musée Royal de l'Afrique Centrale; 1962. p. 81–105.

Kretzoi M, Dobosi VT. Vértesszőllős Site. Akadémiai Kiadó, Budapest: Man and Culture; 1990.

Latham AG, Herries AIR. The formation and sedimentary infilling of the Cave of Hearths and historic cave complex, Makapansgat, South Africa. Geoarchaeology. 2004;19:323–42.

Le Grand Y Approche méthodologique et technologique d'un site d'habitat du Pléistocène moyen: La grotte n. 1 du Mas de Caves (Lunel Viel, Hérault). Unpublished Ph.D. thesis. Provence: Université de Provence; 1994.

Leakey MD. Olduvai Gorge: excavations in beds I and II, 1960–1963. Cambridge: Cambridge University Press; 1971.

Lebel S, Trinkaus E. Middle Pleistocene human remains from Bau de l'Aubesier. J Hum Evol. 2002;43:659–85.

Leesch, D. Hauterives-Champréveyres, 10. Un Campement Magdalénien au Bord du Lac de Neuchâtel: Cadre Chronologique et Culturel, Mobilier et Structures, Analyse Spatiale (Secteur 1). Neuchâtel: Service et Musée Cantonal d'Archéologie; 1997.

Leesch D, Bullinger J, Cattin M-I, Müller W, Plumettaz N. Hearths and hearth-related activities in Magdalenian open-air sites: the case studies of Champréveyres and Monruz (Switzerland) and their relevance to an understanding of Upper Paleolithic site structure. Paper presented at the Magdalenian in Central Europe: New Finds and Concepts. 20–23 September 2005, Poland: Rzeszów; 2005.

Leroi-Gourhan A. Dictionnaire de la préhistoire. Paris: Presses Universitaires de France; 1988.

Leroi-Gourhan A, Brézillon M. Fouilles de Pincevent: Essai d'Analyse Ethnographique d'un Habitat Magdalénien. Paris: Editions du CNRS; 1972.

Lewis HT. The role of fire in the domestication of plants and animals in Southwest Asia: a hypothesis. Man. 1972;7:195–222.

Lewis HT. Comments on "Hominid use of fire in the Lower and Middle Pleistocene: review of the evidence" (James, S.R). Curr Anthropol. 1989;30:14–5.

Linford NT, Canti MG. Geophysical evidence for fires in antiquity: preliminary results from an experimental study. Paper presented at the EGS XXIV general assembly in the Hague, April 1999. Archaeol Prospection. 2001;8:211–25.

Ljubin VP, Bosinski G. The earliest occupation of the Caucasus region. In: Roebroeks W, van Kolfschoten T, editors. The earliest occupation of Europe. Leiden: University of Leiden; 1995. p. 207–53.

Lucretius. On the nature of things. Oxford: Clarendon. (Translation by C. Bailey); 1910.

Madsen B, Goren-Inbar N. Acheulian giant core technology and beyond: an archaeological and experimental case study. Eurasian Prehist. 2004;2(1):3–52.

Mallol C, Marlowe FW, Wood BM, Porter CC. Earth, wind, and fire: ethnoarchaeological signals of Hadza fires. J Archaeol Sci. 2007;34:2035–52.

Mania D. The earliest occupation of Europe: the Elbe-Saale region (Germany). In: Roebroeks W, van Kolfschoten T, editors. The earliest occupation of Europe. Leiden: University of Leiden; 1995. p. 85–101.

Manzi G. Human evolution at the Matuyama-Brunhes boundary. Evol Anthropol. 2004;13:11–24.

Marder O, Khalaily H, Rabinovich R, Gvirtzman G, Wieder M, Porat N, et al. The Lower Paleolithic site of Revadim quarry: preliminary finds. J Isr Prehist Soc. 1998;28:21–53.

Marlowe FW. Hunter-gatherers and human evolution. Evol Anthropol. 2005;14:54–67.

Mason R. Prehistory of the Transvaal. Johannesburg: Witwatersrand University Press; 1969.

McGrew WC. Comments on "Hominid use of fire in the Lower and Middle Pleistocene: review of the evidence" (James, S.R). Curr Anthropol. 1989;30:16–7.

Meignen L, editor. L'abri des Canalettes. Un Habitat Moustérien sur les Grands Causses (Nant, Aveyron). Fouilles 1980–1986. Paris: Editions CNRS; 1993.

Meignen L, Goldberg P, Bar-Yosef O. Chapter 5: the hearths at Kebara Cave and their role in site formation processes. In: Bar-Yosef O. Meignen L, editors. Kebara Cave, Mt. Carmel, Israel: The Middle and Upper Paleolithic archaeology Part I. Harvard: Peapody Museum Press; 2007. p. 91–122.

Melamed Y. Reconstruction of the Hula Valley vegetation and the hominid vegetarian diet by the Lower Palaeolithic botanical remains from Gesher Benot Ya'aqov. Unpublished Ph. D. thesis (Hebrew), Bar-Ilan University, Ramat-Gan; 2003.

Melamed Y. Reconstruction of the landscape and the vegetarian diet at Gesher Benot Ya'aqov archaeological site in the Lower Palaeolithic period. Unpublished M.Sc. thesis (Hebrew), Ramat-Gan: Bar-Ilan University; 1997.

Mellars P. Fire ecology, animal populations and man: a study of some ecological relationships in prehistory. Proc Prehist Soc. 1976; 42:15–45.

Melson WG, Potts R. Origin of reddened and melted zones in Pleistocene sediments of the Olorgesailie Basin, Southern Kenya Rift. J Archaeol Sci. 2002;29:307–16.

Mercier N, Valladas H, Valladas G, Reyss J-L, Jelink A, Meignen L, et al. TL dates of burnt flints from Jelinek's excavations at Tabun and their implications. J Archaeol Sci. 1995;22:495–509.

Mercier N, Froget L, Mallier D, Pilleyre T, Sanzelle S, Tribolo C. Nouvelles données chronologiques pour le site de Menez-Dregan 1 (Bretagne): l'apport de la thermoluminescence. Quaternaire. 2004; 15(3):253–61.

Metcalfe D, Heath KM. Microrefuse and site structure: the hearths and floors of the Heartbreak Hotel. Am Antiquity. 1990;55:781–96.

Meyerhof W. Elucidation of mammalian bitter taste. Rev Physiol Biochem Pharmacol. 2005;154:37–72.

Milton K. A hypothesis to explain the role of meat-eating in human evolution. Evol Anthropol. 1999;8:11–21.

Moigne AM, Barsky DR. Large mammal assembalges from Lower Palaeolithic sites in France: la Caune de l'Arago, Terra Amata, Orgnac 3 and Cagny l'Epinette. In: Gaudzinski S, Turner E, editors. The role of early humans in the accumulations of Lower and Middle Paleolithic bone assemblages. Mainz: Römisch-Germanisches Zentralmuseum; 1999. p. 219–235.

Moncel MH, Moigne AM, Combier J. Pre-Neandertal behavior during isotopic stage 9 and the beginning of 8. New data concenring fauna and lithics in the different occupation levels of Orgnac 3 (Ardèche, South-East France): occupation types. J Archaeol Sci. 2005;32:1283–301.

Monchot H. Les Petits Bovidés de la Caune de l'Arago (Tautavel, Pyrenées-Orientales). Unpublished Ph.D. thesis. Marseille: Université Aix-Marseille II; 1996.

Monnier JL, Hallegouet B, Hinguant S, Laurent M, Auguste P, Bahain JJ, Falguères C, Gebhardt A, Marguérie D, Molines N., Morzadec H, Yokoyama Y. A new regional group of the Lower Paleolithic in Brittany (France), recently dated by Electron Spin Resonance. Paris: Comptes Rendus de l'Académie des Sciences, 319(II); 1994. p. 155–60.

Morton AGT. Archaeological site formation: experiments in lake margin processes. Unpublished Ph.D. thesis. Cambridge: Cambridge University; 1995.

Movius HL. A wooden spear of third Interglacial age from Lower Saxony. Southwest J Anthropol. 1950;6:139–42.

Nadel D. Indoor/outdoor flint knapping and minute debitage remains: the evidence from the Ohalo II submerged camp (19.5 KY, Jordan Valley). Lithic Technol. 2001;26:118–37.

Naveh Z. Ancient man's impact on the Mediterranean landscape in Israel – ecological and evolutionary perspectives. In: Bottema S, Entjes-Nieborg G, Van-Zeist W, editors. Man's role in the shaping of the Mediterranean landscape. Rotterdam: A.A. Balkema; 1990. p. 43–50.

Naveh Z, Carmel Y. The evolution of the cultural landscape in Israel as affected by fire, grazing, and human activities. In: Wasser SP, editor. Evolutionary theory and processes: modern horizons. Papers in honour of Eviatar Nevo. Dordrecht: Kluwer; 2004. p. 337–409.

Neuville R. Le Paléolithique et le Mésolithique du Désert de Judée. Paris: Masson et CIE; 1951.

Nicholson RA. A morphological investigation of burnt animal bone and an evaluation of its utility in archaeology. J Archaeol Sci. 1993;20:411–28.

Oakley KP. Evidence of fire in South African cave deposits. Nature. 1954;174:261–2.

Oakley KP. Fire as Palaeolithic tool and weapon. Proc Prehist Soc. 1956;21:36–48.

Oakley KP. On man's use of fire, with comments on tool-making and hunting. In: Washburn SL, editor. Social life of early man. Chicago, IL: Aldine; 1961. p. 176–93.

O'Connell JF. Alyawara site structure and its archaeological implications. Am Antiquity. 1987;52:74–108.

O'Connell JF, Hawkes K, Blutron-Jones NG. Grandmothering and the evolution of Homo erectus. J Hum Evol. 1999;36:461–85.

Olive M, Taborin Y, editors. Nature et Fonction des Foyers Préhistoriques. Nemours: Association Pour la Promotion de la Recherche Archéologique en Ile de France; 1989.

Oms O, Parés JM, Martínez-Navarro B, Agustí J, Toro I, Martínez-Fernández G, et al. Early human occupation of Western Europe: paleomagnetic dates for two Paleolithic sites in Spain. PNAS. 2000;97:10666–70.

Parés JM, Pérez-González A. Magnetochronology and stratigraphy at Gran Dolina section, Atapuerca (Burgos, Spain). J Hum Evol. 1999;37:325–42.

Parfitt SA, Barendregt RW, Breda M, Candy I, Collins MJ, Coope GR, et al. The earliest record of human activity in northern Europe. Nature. 2005;438:1008–12.

Pastó I, Allué E, Vallverdú J. Mousterian hearths at Abric Romaní. In: Stringer CB, Barton RNE, Finlayson JC, editors. Neanderthals on the edge. Oxford: Oxbow Books; 2000. p. 59–67.

Patel T. Burnt stones and rhino bones hint at earliest fire. New Sci. 1995;1982:5.

Pécsi M. Geomorphological position and absolute age of the Vértesszőllős Lower Palaeolithic site. In: Kretzoi M, Dobosi VT, editors. Vértesszőllős site, man and culture. Budapest: Akadémiai Kiadó; 1990. p. 27–41.

Perlès C. Préhistoire du Feu. Paris: Masson; 1977.

Picard L. The Quaternary in the Northern Jordan Valley. Proc Isr Acad Sci Humanit. 1963;1(4):1–34.

Picard L. The Pleistocene peat of Lake Hula. Bulletin of the Research Council of Israel II(2); 1952. p. 147–156.

Pickering TR, Egeland CP, Domínguez-Rodrigo M, Brain CK. Schnell AG. Testing the "shift in the balance of power" hypothesis at. Swartkrans, South Africa: Hominid cave use and subsistence behavior. in the Early Pleistocene. J Anthropol Archaeol. 2008;27:30–45.

Pliny. Natural history. Cambridge: Harvard University Press. (Translation by D.E. Eichholz, The Loeb Classical Library); 1962.

Plumettaz N. Le Site Magdalénien de Monruz, 2: Étude des Foyers à Partir de l'Analyse des Pierres et de Leurs Remontages. Neuchâtel: Office et Musée Cantonal d'Archéologie; 2007. Plumettaz N. Les structures de combustion. In: BullingerJ, Leesch D.

Plumettaz N, editors. Le Site Magdalénien de Monruz 1: Premiers Éléments pour l'Analyse d'un Habitat de Plein Air. Neuchâtel: Service et Musée Cantonal d'Archéologie (Archéologie Neuchâteloise 33); 2006. p. 45–65.

Plummer T. Flaked stones and old bones: biological and cultural evolution at the dawn of technology. Yearbook Phys Anthropol. 2004;47:118–64.

Pope GG. Evidence on the age of the Asian Hominidae. PNAS. 1983; 80:4988–92.

Potts R. Home bases and early hominids. Am Sci. 1984;72:338–47.

Potts R. Variables versus models of early Pleistocene hominid land use. J Hum Evol. 1994;27:7–24.

Preece RC, Gowlett JAJ, Parfitt SA, Bridgland DR, Lewis SG. Humans in the Hoxnian: habitat, context and fire use at Beeches Pit, West. Stow, Suffolk, UK. J Quatern Sci. 2006;21:485–96.

Preece RC, Parfitt SA, Bridgland DR, Lewis SG, Rowe PJ, Atkinson TC, et al. Terrestrial environments during MIS 11: evidence from the Palaeolithic site at West Stow, Suffolk, UK. Quatern Sci Rev. 2007;26:1236–300.

Purdy BA. Fractures for the archaeologist. In: Swanson E, editor. Lithic technology: making and using stone tools. Paris: Mouton; 1975. p. 133–41.

Purdy BA. Pyrotechnology: prehistoric application to chert materials in North America. In: Wertime TA, Wertime SF, editors. Early pyrotechnology: the evolution of the first fire-using industries. Washington, DC: Smithsonian Institution Press; 1982. p. 31–44.

Purdy BA, Brooks HK. Thermal alteration of silica minerals: an archeological approach. Science. 1971;173:322–5.

Pyne SJ, Andrews PL, Laven RD. Introduction to wildland fire. New York: Wiley; 1996.

Rabinovich R, Gaudzinski-Windheuser S, Goren-Inbar N. Systematic butchering of fallow deer (*Dama*) at the Early Middle Pleistocene Acheulian site of Gesher Benot Ya'aqov (Israel). J Hum Evol. 2008;54:134–49.

Rabinovich R, Gaudzinski-Windheuser S, Kindler L, Goren-Inbar N. The Acheulian Site of Gesher Benot Ya'aqov Volume III: Mammal Taphonomy – The Assemblages of Layers V-5 and V-6. Dordrecht: Springer; in press.

Raposo L, Santonja M. The earliest occupation of Europe: the Iberian peninsula. In: Roebroeks W, van Kolfschoten T, editors. The earliest occupation of Europe. Leiden: University of Leiden; 1995. p. 7–25.

Rathje WL. Modern material culture studies. Adv Archaeol Meth Theor. 1979;2:1–37.

Raum OF. The social functions of avoidances and taboos among the Zulu. Berlin: de Gruyter; 1973.

Raynal L-P, Magoga L, Bindon P. The tephrofacts and the first human occupation of the French Massif Central. In: Roebroeks W, van Kolfschoten T, editors. The earliest occupation of Europe. Leiden: University of Leiden; 1995. p. 129–46.

Rigaud J-P, Geneste JM. L'utilisation de l'espace dans la grotte Vaufrey. In: Rigaud J-P, editor. La Grotte Vaufrey. Mémoires de la Société Préhistorique Française 19; 1988. p. 593–611.

Rigaud J-P, Simek JF, Ge T. Mousterian Fires from Grotte XVI (Dordogne, France). Antiquity. 1995;69:902–12.

Rink WJ, Schwarcz HP. Short contribution: ESR and Uranium Series dating of teeth from the Lower Paleolithic site of Gesher Benot. Ya'aqov, Israel: confirmation of Paleomagnetic age indications. Geoarchaeology. 2005;20:57–66.

Rink WJ, Schwarcz HP, Ronen A, Tsatskin A. Confirmation of a near 400 ka age for the Yabrudian industry at Tabun Cave, Israel. J Archaeol Sci. 2004;31:15–20.

Roberts MB, Parfitt SA. Boxgrove: a Middle Pleistocene hominid site at Eartham Quarry, Boxgrove, West Sussex. London: English Heritage; 1999.

Roe DA. The Kalambo Falls large cutting tools: a comparative metrical and statistical analysis. In: Clark JD, editor. Kalambo Falls prehistoric site Volume III – the earlier cultures: Middle and Earlier Stone Age. Cambridge: Cambridge University Press; 2001. p. 429–599.

Roebroeks W. Hominid behaviour and the earliest occupation of Europe: an exploration. J Hum Evol. 2001;41:437–61.

Roebroeks W. From find scatter to early hominid behaviour: a study of Middle Paleolithic riverside settlements at Maastricht-Belvédère (The Netherlands). Analecta Praehistorica Leidensia 21, Leiden: Leiden University Press; 1988.

Roebroeks W, van Kolfschoten T. The earliest occupation of Europe. Leiden: University of Leiden; 1995.

Rolland N. Was the emergence of home bases and domestic fire a punctuated event? a review of the Middle Pleistocene record in Eurasia. Asian Perspect. 2004;34:248–80.

Rolland N. Cave occupation, fire-making, hominid/carnivore coevolution, and Middle Pleistocene emergence of home-base settlement systems. Acta Anthropologica Sinica 2000;19 Suppl:209–17.

Ronen A, Burdukiewicz J-M, Laukhin SA, Winter Y, Tsatskin A, Dayan T, et al. The Lower Palaeolithic site Bizat Ruhama in the Northern Negev, Israel: preliminary report, 1996 excavations. Archaeologisches Korrespondenzblatt. 1998;28:163–73.

Rose L, Marshall F. Meat eating, hominid sociality, and home bases revisited. Curr Anthropol. 1996;37:307–38.

Rowlett RM. Fire control by Homo erectus in East Africa and Asia. Acta Anthropologica Sinica 2000; 19 Suppl:198–208.

Rowlett RM, Davis MG, Graber RB. Friendly fire. Discovering Archaeol. 1999;1(5):82–9.

Rudgley R. The lost civilizations of the Stone Age. New York: The Free Press; 1999.

Sanchis Serra A, Fernández-Peris J. Procesado y consumo antrópico de conejo en la Cova del Bolomor (Tavernes de la Valldigna, Valencia): El nivel XVIIc (ca 350 ka). Complutum. 2008; 19(1):25–46.

Santonja M, Villa P. The Acheulian of Western Europe. In: Goren-Inbar N, Sharon G, editors. Axe age: Acheulian toolmaking from quarry to discard. London: Equinox; 2006. p. 429–78.

Saragusti I, Goren-Inbar N. The biface assemblage from Gesher Benot Ya'aqov, Israel: illuminating patterns in "Out of Africa" dispersal. Quatern Int. 2001;75:85–9.

Sauer CO. The agency of man on the earth. In: Thomas WL, editor. Man's role in changing the face of the Earth. Chicago, IL: The University of Chicago Press; 1956. p. 49–69.

Sauer CO. Sedentary and mobile bents in early societies. In: Washburn SL, editor. Social life of early man. Chicago, IL: Aldine; 1961. p. 256–66.

Sauer CO. Early relations of man to plants. In: Leighly J, editor. Land and life: a selection from the writings of Carl Ortwin Sauer. Berkeley, CA: University of California Press; 1947 (1963). p. 155–81. (Originally published in Geogr Rev 1947;37:1–25).

Sauer CO. Fire and early man. In: Leighly J, editor. Land and life: a selection from the writings of Carl Ortwin Sauer. Berkeley: University of California Press; 1961b (1963). p. 288–99.

Scarre C. Exploring prehistoric Europe. Oxford: Oxford University Press; 1998.

Schiegl S, Lev-Yadun S, Bar-Yosef O, El-Goresy A, Weiner S. Siliceous aggregates from prehistoric wood ash: a major component of sediments in Kebara and Hayonim caves (Israel). Isr J Earth Sci. 1994;43:267–78.

Schiegl S, Goldberg P, Bar-Yosef O, Weiner S. Ash deposits in Hayonim and Kebara Caves, Israel: macroscopic, microscopic and mineralogical observations, and their archaeological implications. J Archaeol Sci. 1996;23:763–81.

Schiffer MB. Archaeological context and systemic context. Am Antiquity. 1972;37:156–65.

Schiffer MB. Behavioral archeology. New York: Academic; 1976.

Schiffer MB. Formation processes of the archaeological record. Albuquerque: University of New Mexico Press; 1987.

Scott AC, Moore J, Brayshay B. Introduction to fire and the palaeoenviroment.

Palaeogeogr Palaeoclimatol Palaeoecol 2000; 164:vii–xi.

Sept JM. Was there no place like home? Curr Anthropol. 1992; 33:187–207.

Sergant J, Crombé P, Perdaen Y. The "invisible" hearths: a contribution to the discernment of Mesolithic non-structured surface hearths J Archaeol Sci. 2006;33:999–1007.

Shahack-Gross R, Bar-Yosef O, Weiner S. Black-coloured bones in Hayonim Cave, Israel: differentiating between burning and oxide staining. J Archaeol Sci. 1997;24:439–46.

Sharon G. Acheulian large flake industries: technology, chronology, and significance. BAR International Series. Volume 1701. Oxford: Archaeopress; 2007.

Sharon G, Goren-Inbar N. Soft percussor use at the Gesher Benot Ya'aqov Acheulian site? J Isr Prehist Soc. 1999;28:55–79.

Shea JJ. Artifact abrasion, fluvial processes, and "living floors" from the Early Paleolithic site of 'Ubeidiya (Jordan Valley, Israel). Geoarchaeology. 1999;14:191–207.

Shepherd R. Mining in Europe during the Neolithic and the Chalcolithic. In: De Laet SJ, editor. History of humanity Volume I: prehistory and the beginnings of civilization. London: Routledge; 1994. p. 616–26.

Shipman P, Foster GF, Scheninger M. Burnt bones and teeth: an experimental study of colour, morphology, crystal structure and shrinkage. J Archaeol Sci. 1984;11:307–25.

Silverman BW. Density estimation for statistics and data analysis. New York: Chapman & Hall; 1986.

Simek JF. Integrating pattern and context in spatial archaeology. J Archaeol Sci. 1984;11:405–20.

Simms SR. The archaeological structure of a Bedouin camp. J Archaeol Sci. 1988;15:197–211.

Skinner A, Lloyd J, Brain C. Thackeray F. Electron spin resonance and the first use of fire. Paper presented at the 69th Annual Meeting of the Paleoanthropology Society and Society for American Archaeology. March 30–April 4, Montreal; 2004.

Spurling B, Hayden B. Ethnoarchaeology and intrasite spatial analysis: A case study from the Australian Western desert. In: Hietala HJ, editor. Intrasite spatial analysis in archaeology. Cambridge: Cambridge University Press; 1984. p. 224–41.

Stahl AB. Hominid dietary selection before fire. Curr Anthropol. 1984;25(2):151–68.

Stapert D. Middle Palaeolithic dwellings: fact or fiction? Some applications of the ring and sector method. Palaeohistoria. 1990;32:1–19.

Stein JK, Teltser PA. Size distributions of artifact classes: combining macro-and-micro-fractions. Geoarchaeology. 1989;4:1–30.

Stekelis M. The Palaeolithic deposits of Jisr Banat Ya'qub. Bull Res Counc Isr. Geo-Sci. 1960;9G(2–3):61–90.

Stevenson MG. Beyond the formation of hearth-associated artifact assemblages. In: Kroll EM, Price TD, editors. The Interpretation. of archaeological spatial patterning. New York: Plenum; 1991. p. 269–99

Stewart OC. Fire as the first great force employed by man. In: Thomas WL, editor. Man's role in changing the face of the Earth. Chicago. IL: The University of Chicago Press; 1956. p. 115–33.

Stiner MC, Kuhn SL, Weiner S, Bar-Yosef O. Differential Burning. recrystallization, and fragmentation of archaeological bone. J Archaeol Sci. 1995;22:223–37.

Stiner MC, Barkai R, Gopher A. The Acheulo-Yabrudian faunas from Qesem Cave (Israel): preliminary results. Paper presented at the. 69th Annual Meeting of the Paleoanthropology Society and Society for American Archaeology. March 30–April 4, Montreal; 2004.

Tankersley KB. Clovis mastic and its hafting implications. J Archaeol Sci. 1994;21:117–24.

Texier PJ, Brugal JP, Lemorini C, Wilson L. Fonction d'un site du. Paléolithique moyen en marge d'un territoire: l'abri de La Combette (Bonnieux, Vaucluse). In: Brugal JP, Meignen L, Patou-Mathis M editors. Economie Préhistorique: les Comportements de Subsistance au Paléolithique. Sophia Antipolis: Editions APDCA; 1998. p. 325–48.

Théry I, Gril J, Vernet JL, Meignen L, Maury J. Coal used for fuel at two prehistoric sites in Southern France: les Canalettes (Mousterian) and les Usclades (Mesolithic). J Archaeol Sci. 1996;23:509–12.

Théry-Parisot I. Économie des Combustibles au Paléolithique. Paris: CNRS; 2001.

Thevenon F, Anselmetti FS. Charcoal and fly-ash particles from Lake Lucerne sediments (Central Switzerland) characterized by image analysis: anthropologic, stratigraphic and environmental implications. Quatern Sci Rev. 2007;26:2631–43.

Thieme H. Lower Paleolithic hunting spears from Germany. Nature. 1997;385:807–10.

Thieme H. Lower Palaeolithic hunting weapons from Schöningen, Germany: the oldest spears in the world. Acta Anthropologica. Sinica 2000;19 Suppl:136–43.

Thieme H, Viel S. Neue untersuchungen zum eemzeitlichen Elefanten-Jagdplatz Lehringen. Die Kunde N. F. 1985;36:11–58.

Thoma A. Human tooth and bone remains from Vértesszőllős. In: Kretzoi M, Dobosi VT, editors. Vértesszőllős site, man and culture. Budapest: Akadémiai Kiadó; 1990. p. 253–62.

Thomas WL, editor. Man's role in changing the face of the Earth. Chicago, IL: The University of Chicago Press; 1956.

Tuffreau A, Sommé J, editors. Le Gisement Paléolithique Moyen de Biache-Saint-Vaast (Pas de Calais). Mémoires de la Société Préhistorique Française 1988;21:1–340.

Ungar PS, Grine FE, Teaford MF. Diet in early Homo: a review of the evidence and a new model of adaptive versatility. Ann Rev Anthropol. 2006;35:209–28.

Valladas H, Mercier N, Joron J-L, Reyss J-L. GIF laboratory dates for Middle Paleolithic Levant. In: Akazawa T, Aoki K, Bar-Yosef O, editors. Neandertals and modern humans in Western Asia. New York: Plenum; 1998. p. 69–75.

Valleverdú J, Courty MA, Carbonell E, Canals A, Burjachs F. Les sédiments *d'homo antecessor* de Gran Dolina, Sierra de Atapuerca, Burgos, Espagne: Interprétation micromorphologique des processus de formation et enregistrement paléoenvironmental des sédiments. L'Anthropologie. 2001;105:45–69.

Valoch K. The earliest occupation of Europe: Eastern Central and Southeastern Europe. In: Roebroeks W, van Kolfschoten T, editors. The earliest occupation of Europe. Leiden: University of Leiden; 1995. p. 67–84.

Vance ED. Microdebitage and archaeological activity analysis. Archaeology. 1987;40(4):58–9.

Vaquero M, Pastó I. The definition of spatial units in Middle Palaeolithic sites: the hearth-related assemblages. J Archaeol Sci. 2001;28:1209–20.

Vega Toscano LG, Raposo L, Santonja M. Environments and settlements in the Middle Palaeolithic of the Iberian Peninsula. In: Roebroeks W, Gamble C, editors. The Middle Palaeolithic occupation of Europe. Leiden: University of Leiden; 1999. p. 23–48.

Vertes L, Dobosi VT. Fireplaces of the settlement. In: Kretzoi M, Dobosi VT, editors. Vértesszőllős site, man and culture. Budapest: Akadémiai Kiadó; 1990. p. 519–21.

Villa P. Terra Amata and the Middle Pleistocene archaeological record of Southern France. California: University of California Press; 1983.

Villa P. Middle Pleistocene prehistory in southwestern Europe: the state of our knowledge and ignorance. J Anthropol Res. 1991;47: 193–218.

Villa P. Europe: Lower and Middle Pleistocene archaeology. In: De Laet SJ, editor. History of humanity Volume I: Prehistory and the beginnings of civilization. London: Routledge; 1994. p. 44–61.

Villa P. Fire and fireplaces in the Middle and Upper Pleistocene of Western Europe. Paper presented at the XIVth UISPP Congress, 2–8 September 2001, Liege, Belgium; 2001.

Villa P, Bon F. Fire and fireplaces in the Lower, Middle and early Upper Paleolithic of Western Europe. J Hum Evol. 2002;42:A37.

Villa P, Lenoir M. Hunting and hunting weapons of the Lower and Middle Paleolithic of Europe. In: Richards M, Hublin JJ, editors. The evolution of hominid diet: integrating approaches to the study of Palaeolithic subsistence. Dordrecht: Springer; in press. p. 59–84.

Villa P, Bon F, Castel J-C. Fuel, fire and fireplaces in the Paleolithic of Western Europe. Rev Archaeol. 2002;23(1):33–42.

Villa P, Castel J-C, Beauval C, Bourdillat V, Goldberg P. Human and carnivore sites in the European Middle and Upper Paleolithic:. similarities and differences in bone modification and fragmentation. Rev Paléobiol. 2004;23(2):705–30.

Voytek B. Some theoretical issues connected with the study of filnt mining. In: Schild R, Sulgostowska Z editors. Man and flint: proceedings of the VIIth international flint symposium Warszawa – Ostrowiec S´ wie̜tokrzyski, September 1995. Warszawa: Institute of Archaeology and Ethnology Polish Academy of Sciences; 1997. p. 149–52.

Wadley L. Putting ochre to the test: replication studies of adhesives that may have been used for hafting tools in the Middle Stone Age. J Hum Evol. 2005;49:587–601.

Wandsnider L. The roasted and the boiled: food composition and heat treatment with special emphasis on pit-hearth cooking. J Anthropol. Archaeol. 1997;16:1–48.

Wang X, Thomas SD, Zhang J. Relaxation of selective constraint and loss of function in the evolution of human bitter taste receptor genes. Hum Mol Genet. 2004;13:2671–78.

Weiner S, Xu Q, Goldberg P, Liu J, Bar-Yosef O. Evidence for the use. of fire at Zhoukoudian, China. Science. 1998;281: 251–3.

Weiner S, Goldberg P, Bar-Yosef O. Three-dimensional distribution of minerals in the sediments of Hayonim Cave, Israel: diagenetic processes and archaeological implications. J Archaeol Sci. 2002;29: 1289–308.

Weinstein-Evron M, Bar-Oz G, Zaidner Y, Tsatskin A, Druck D, Porat N, et al. Introducing Misliya Cave, Mount Carmel, Israel: a new continuous Lower/Middle Paleolithic sequence in the Levant. Eurasian Prehist. 2003;1(1):31–55.

Westbroek P, Collins MJ, Jansen JHF, Talbot LM. World archaeology and global change: did our ancestors ignite the Ice Age? World Archaeol. 1993;25:122–33.

Whelan RJ. The ecology of fire. Cambridge: Cambridge University Press; 1995.

White MJ. The Clactonian question: on the interpretation of core-and-flake assemblages in the British Lower Paleolithic. J World Prehist. 2000;14:1–63.

Wobber V, Hare B, Wrangham R. Great apes prefer cooked food. J Hum Evol. 2008;55:340–8.

Wrangham RW, Jones JH, Laden G, Pilbeam D, Conklin-Brittain N. The raw and the stolen: cooking and the ecology of human origins. Curr Anthropol. 1999;40:567–94.

Wymer J. The Lower Palaeolithic occupation of Britain. Salisbury: Wessex Archaeology/English Heritage; 1999.

Yellen JE. Archaeological approaches to the present: models for reconstructing the past. New York: Academic; 1977.

Zaidner Y, Ronen A, Burdukiewicz J-M. L'industrie microlithique du Paléolithique inférieur de Bizat Ruhama, Israël. L'Anthropologie. 2003;107:203–22.

Subject Index

A

Acheulian, 1, 3–6, 8, 11, 13, 14, 19, 75, 79, 83, 85, 88, 90–93, 95–97
Acheulo-Yabrudian, 5
Adhesives. *See* Hafting
Apes, chimpanzees, 93

B

Basalt
  cleavers (*see* Bifacial tools)
  cores, giant cores, 16, 87, 89
  dominance of basalt, 84
  handaxes (*see* Bifacial tools)
  spatial distribution of basalt, 84
  weathering, 89
Bifacial tools
  cleavers, 16, 26, 33, 84, 87–90
  experimental studies (*see* Experimental studies)
  handaxes, 16, 26, 33, 84, 87–90
  homogeneity, 89
  products of handaxe manufacture, 84
  standardization, 89
  stocking, 89, 90, 92
  technological analyses, 89
  workshop-site of, 88–89
Bitter taste. *See* Diet
Bones
  burned, 4–8, 10
  identification of burning
    discoloration, 10, 11, 21–23
    fragmentation, 10, 11, 22, 23
  preservation, 10
  spatial distribution of, 86
  used as fuel, 7

C

Charcoal
  microscopic charcoal (*see* Fire, ecology)
Cut marks, 16, 85, 86, 90

D

Diet
  bitter taste, 93
  consumption of meat, 92, 93
  consumption of plant food, 93
  cooking/roasting
    cooked food, 93
    experimental studies, 92
    as means of detoxifying poisonous foodstuffs, 93
    preference for cooked food, 93
  expensive-tissue hypothesis, 92
  gut size, 92
  nonsomatic digestion, 92
  nuts (*see* Nuts)
Discard patterns
  hearth-related, 20
  small refuse, 20
Drop zone, 20

E

Electron Spin Resonance (ESR), 4, 5, 7, 10, 13
Elephant
  at Gesher Benot Ya'aqov (*Paleoloxodon antiquus*)
    battered elephant skull, 16
  at Lehringen, 94
  at Torralba, 8, 94
Ethnography, 1, 19–21, 93, 94, 96
Eurasia, 14, 97, 98
Europe
  colonization of, 14, 98
  earliest occupation, 98
Expensive-tissue hypothesis. *See* Diet
Experimental studies
  adhesives, 94
  consumption of cooked meat (*see* Diet)
  effects of fire on flint
    fragmentation, 23
    heating of flint, 23
  experimental fireplace, 2, 7
  fire ignition (*see* Fire, ecology)
  knapping at GBY
    basalt handaxes and cleavers, 90
  lake margin processes, 22, 23, 73
  vertical variation of temperatures in a fire (*see* Fire, ecology)

F

Fallow deer (*Dama*)
  butchering of, 85, 92
  cut marks, 85
  marrow extraction, 85
  percussion marks, 85
  spatial distribution of, 85
Fire
  African tradition, 97
  anthropogenic
    fire and human activities at GBY, 96
    hearth-related spatial patterning, 96
    spatial patterns, 24, 75, 83
  behavioral and technological aspects
    fire as a process of civilization, 95

Fire (*Con't.*)
fire in food preparation (*see* Diet)
fire-keeping, fire-keeper, 94
fire-making, fire-maker, 94, 97
fire used for hunting (*see* Hunting)
magnitude of the use of fire, 97
social role of fire, 94
early evidence
Africa, 4
Asia, 5–6
Europe, 6–8
Levant, 4–5
ecology
cold lightning, 76
combustion, 75–76
fire behavior, 77
fire dynamics, 75
fire history, 75
fires in waterlogged environments, 77
fuel moisture, 76, 77
heat transfer, 76–77
hot lightning, 76
ignition, 75–76
lightning, 75–78
microscopic charcoal, 75
probability of ignition, 76
spontaneous ignition, 76
subsurface fire, 78–79
surface fire, 78
vertical variation of temperatures, 77
wildfire, 76–78
evolutionary incentive
to *Homo erectus*, 92
identification of
bones (*see* Bones)
effects of fire on flint, 77, 78
sediments (*see* Sediments)
stones, 10
natural fires
natural fires, 75, 78, 79
wildfires (*see* Fire, ecology)
Fireplace, 2, 3, 7–10. *See also* Hearth
Flint
controlled heating of, 94
effects of fire on, 22–24, 77, 78
identification of burning
crazing, 23
discoloration, 23
experimental studies (*see* Experimental studies)
fragmentation, 23
potlid fractures, 23
predominance of, 84
Food-sharing hypothesis. *See* Home base, central-place foraging
Fuel
bones used as fuel, 7
fuel moisture (*see* Fire, ecology)
lichen used as fuel, 10
lignite used as fuel, 10

G
Geographic Information System (GIS)
ArcMap, 28
ArcScene, 28

H
Hafting
adhesives, 94
experimental studies, 94
Handaxes. *See* Bifacial tools
Hearth
in dwelling structures, 8
hearth and home (*see* Home and fire)
hearth-related discard patterns, 20
hearth-related spatial patterning, 19–22, 96
interior hearths (Terra Amata), 8
Home and fire, 95
Home base, central-place foraging, 95
Homo erectus, 2, 6–8, 92, 97
anatomical traits of, 92
Human dispersal, out of Africa. *See* Out of Africa
Human migration, out of Africa. *See* Out of Africa
Hunting
use of fire in hunting, 93
with wooden spears (*see* Wood, spears)
Hydraulic jumble, 23
Hydrodynamics. *See* Lake margin processes

L
Lake margin processes, 22, 23, 73
hydrodynamics, 23
Lake shore environment, 22
Levantine Corridor, 1, 14, 97
Lightning. *See* Fire, ecology
Lithics
basalt (*see* Basalt)
cleavers (*see* Bifacial tools)
experimental studies on burning (*see* Experimental studies)
flint (*see* Flint)
hafting (*see* Hafting)
handaxes (*see* Bifacial tools)
Kombewa, 89
Levallois, 89
limestone, 5, 6, 15, 25, 26, 33, 84, 87
mining (*see* Mining)
quarrying (*see* Quarrying)

M
Macroartifacts, 1, 25–28, 33, 34, 36, 39–41, 43, 47, 50, 51, 55, 58, 60, 61, 63, 66, 68, 70, 73, 78, 82, 84–86, 96
McKellar's principle, 20
Mediterranean zone, climate, 76
Microartifacts, 1, 20, 22–31, 33–71, 73, 74, 78, 79, 82–86, 88–92, 96, 97
terminology, 20
Microdebitage, 20
Microflakes, 20
Microchunks, 20
Mining, 93

N
Nuts, 16, 90, 91
nut cracking, 16, 91, 92, 96
pitted anvils (*see* Pitted anvils pitted stones)
processing of, 91, 92, 96

O
Oldowan, 53
Out of Africa, x, 8, 14, 97, 98
fire as a triggering factor, 14, 75, 97
human dispersal, 8
human migration, 8, 14, 97

P
Pebbles, 1, 8, 15, 26, 29, 31, 33, 47, 49, 50, 52, 55, 58, 60, 61, 63, 64, 66, 68, 69, 73, 74, 78, 82

Phantom hearths, 21–22, 88, 90, 96, 97
Phytoliths, 2
Pitted anvils pitted stones
  spatial analyses of, 90
Pollen, 4, 6, 16
Pyrotechnology, 98

Q
Quarrying, 2, 27, 93, 94

R
Residual primary refuse, 20

S
Secondary refuse, 20
Sediments
  ash/ashy sediments, 4, 6, 10
  burned sediments, 2–4, 8–10, 21
  discoloration, 9, 21, 22
  magnetic analyses/paleomagnetic analysis, 3, 6, 10
  mineralogical analyses, 5
  reddened soils, 9
  sedimentological analyses, 10, 11
  wood-ash, 4, 10
Shells, burned, 7, 8
Site structure, 20
Small refuse
  discard patterns, 20
  hearth related spatial patterns (*see* Hearth related spatial patterns)
  microartifacts (*see* Microartifacts)
  spatial patterns, 20
Spatial analyses
  differentiation in distribution of burned and unburned flint, 1, 30, 31, 41, 55, 66, 68, 70, 74, 79–83, 96
  GIS (*see* Geographic Information System)
  hearth-related spatial patterning, 19–22, 79, 96
  homogeneity analysis, 27, 30
  kernel density maps, 29–31, 35, 36, 38, 39, 43, 44, 46, 47, 49–66, 68, 69, 82, 85, 86, 88–91
  patterns of anthropogenic fire, 24, 31, 73
  random patterns of distribution, 31
  search radii, 30
  spatial complexity of activities, 83
  spatial configuration of burned flint, 19, 22, 24
  spatial patterning of activities, 82, 91, 92, 96
  statistical tests
    chi square (χ^2), 31
    standardized residuals (SR), 31
Spatial differentiation of activities. *See* Spatial patterning of activities
Spatial distribution. *See* Spatial analyses
Spatial patterning of activities, 83, 91, 92, 96
Spatial recording, random spatial reference/artificial coordinates, 27
Statistical tests. *See* Spatial analyses
Stocking, 89, 90, 92
Structures latente, 22

T
Taphonomy
  lake margin processes, 23
  taphonomic considerations, GBY, 22–23
  taphonomic problems related to identification of burning, 11
Thermal alteration
  crazing, 11, 23
  discoloration, 3, 5, 11, 21, 23
  effects of fire on flint, 23–24
  fractures, fragmentation, 11, 23
  potlidding, potlid fractures, 3, 11, 23, 24
Thermal demagnetization, 2
Thermoluminescence (TL), on GBY flints, 5
Toss zone, 20

W
Waterlogged sites, 76
Wood
  burned, 11, 15, 16
  hardening, ethnographic evidence, 94
  objects, 8
  spatial distribution of, 85
  spears, hunting with, 94
  taxonomic identification of, 85

Index of Sites

A
Abri Zumoffen, 5
Atapuerca, 98
Azokh (Azych) Cave, 8

B
Barranco Léon, 98
Beeches Pit, 7, 9
Belvédère Quarry, 21, 22
Bezez Cave, 5
Bilzingsleben, 8
Bizat Ruhama, 5
Blassac-les-Battants, 6
Bolomor Cave, 8

C
Cave of Hearths, 2, 4
Champréveyres, 10, 22, 79
Chesowanja, 3, 10
Clacton-on-Sea, 94

F
Fuente Nueva-3, 98

G
Gadeb, 3, 10
Gesher Benot Ya'aqov (GBY)
  animal exploitation and processing, 16
    fallow deer (*Dama*), 85 (*see also* Fallow deer)
  bifacial tools, 13, 15, 16, 26, 84, 88–91
  chronology, 13–14, 97, 98
  depositional sequence, 11–14, 31
  excavation methods and provenance recording, 24–25
  freshwater mollusks (*Melanopsis aaronsohni* and *Viviparus syriacus*), 13
  lithic production, 15–17
  plant exploitation, 16–17
    nuts, nut cracking (*see* Nuts)
  resemblance to African assemblages, 13
  sedimentological analyses, 10, 11, 13, 14
  taphonomic considerations, 22–23
Gongwangling, 6

H
Hoxne, 6

K
Kalambo Falls, 4, 94
Koobi Fora, 2, 3–9, 10

L
La Cotte de St. Brelade, 7
Lantian (Chenjiawo), 6
Latamne, 4, 5
Lehringen, 94

M
Makapansgat, 2
Marks Tey, 6
Menez-Dregan, 7
Middle Awash, 3, 10
Misliya Cave, 5
Monruz, 10, 22, 79
Montières, 6

O
Olduvai Gorge, 95
Olorgesailie, 3, 4

P
Přezletice, 7

Q
Qesem Cave, 5

R
Revadim, 5

S
Šandalja Cave I, 7
San Quirce, 8
Schöningen, 7, 94
Solana del Zamborino, 8
St. Estéve-Janson (Escale Cave), 6
Swanscombe, 6
Swartkrans, 4, 10

T
Tabun Cave, 5
Terra Amata, 8
Torralba, 7, 8, 94
Trinil, 6

U
'Ubeidiya, 4, 14
Umm Qatafa, 5

V
Verrebroek, 22
Vértesszőllős, 7, 10

W
Wonderwerk Cave, 4

X
Xihoudu, 6

Y
Yuanmou, 6

Z
Zhoukoudian, 5, 6, 10

MIX
Papier aus verantwortungsvollen Quellen
Paper from responsible sources
FSC® C105338

If you have any concerns about our products,
you can contact us on
ProductSafety@springernature.com

In case Publisher is established outside the EU,
the EU authorized representative is:
Springer Nature Customer Service Center GmbH
Europaplatz 3, 69115 Heidelberg, Germany

Printed by Libri Plureos GmbH
in Hamburg, Germany